Fundamentals of Heat Exchanger Design

Fundamentals of Heat Exchanger Design

Editor

Shudhkar Mishra

Fundamentals of Heat Exchanger Design

Edited by **Shudhkar Mishra**

Printed in 2017

ISBN: 978-1-68117-377-1

Library of Congress Control Number: 2015941565

© 2016 by

SCITUS Academics LLC,
616, Corporate Way, Suite 2, 4766,
Valley Cottage, NY 10989

www.scitusacademics.com

Contents

Preface

Heat exchangers has increased immensely from the viewpoint of energy conservation, conversion, recovery, and successful implementation of new energy sources. Its importance is also increasing from the standpoint of environmental concerns such as thermal pollution, air pollution, water pollution, and waste disposal. Heat exchangers are used in the process, power, transportation, air-conditioning and refrigeration, cryogenic, heat recovery, alternate fuels, and manufacturing industries, as well as being key components of many industrial products available in the marketplace. The heat exchanger design equation can be used to calculate the required heat transfer surface area for a variety of specified fluids, inlet and outlet temperatures and types and configurations of heat exchangers, including counterflow or parallel flow. A value is needed for the overall heat transfer coefficient for the given heat exchanger, fluids, and temperatures. Heat exchanger calculations could be made for the required heat transfer area, or the rate of heat transfer for a heat exchanger of given area.

Editor

The Design of Heat Exchangers

Arturo Reyes-León[1], Miguel Toledo Velázquez[1], Pedro Quinto-Diez[1], Florencio Sánchez-Silva[1], Juan Abugaber-Francis[1], and Celerino Reséndiz-Rosas[2]

[1]Applied Thermal and Hydraulic Engineering Laboratory SEPI-ESIME-IPN Professional Unit, Lindavista, México D.F.

[2]División de Estudios de Posgrado e Investigación, Instituto Tecnológico de Pachuca, Pachuca de Soto Hidalgo, México

ABSTRACT

A relation between heat transferred and energy loss, for turbulent flow. In different tube arrangements, is made. The conditions are determined which decide the dimensions and velocities for a heat exchanger. Also, a reference to the economic dimensioning of heat exchangers is presented. In this study, the conditions which a heat exchanger must satisfy represent the best balance between the amounts of material employed. The investigation is restricted to the case of turbulent flow.

INTRODUCTION

The heat exchanger is an equipment that allows heat transference between two fluids at different temperatures. Heat exchangers are extensively used in industry due to their wide variety of construction and applications in heat transference processes for producing conventional energy such as condensers, heaters, boilers or steam generators. They provide an adequate surface for heat transference to occur and their mechanical and thermal characteristics allow high pressure and high temperature processes.

Heat exchangers are important, their optimization rises the competitiveness and allows energy saving. The necessity of saving and recovering energy for different processes in industry makes essential the develop of new manufacturing technology for heat exchangers in order to cover a wide range of operation conditions. In recent years, new software for design heat exchangers has been focus in adapting the equipment to the required process and new solutions have been found that make the design time shorter.

HEAT TRANSFER AND FRICTION WORK

Let us first seek a relation between the heat transfer and energy loss. The problem is to determine the energy E which has to be spent in order that a quantity of heat Q may be transferred to a surface F, the average temperature difference being Q [1,2].

Symbols used:

Q = quantity of heat flowing per second.

G = weight of gas or liquid flowing per second.

Q = heat transfer coefficient.

Cp = specific heat at constant pressure.

Δt = temperature variation of medium during flow.

w = velocity of flow.

ζ = coefficient in pressure drop formula

l = tube length.

d = tube diameter.

γ = specific weight.

μ = absolute viscosity.

λ = thermal conductivity.

System units: m, kg, sec.

The heat given up to the surface F is:

$$Q = \alpha \cdot \Theta \cdot F$$

(1)

And the gas loses a corresponding amount of heat

$$Q = G \cdot C_p \cdot \Delta t$$

(2)

Whence

$$\frac{\Delta t}{\Theta} = \frac{\alpha F}{GC_p}$$

(3)

It is convenient to derive the expression of the energy loss first for longitudinal flow through the tubes, and applies it to the case of cross flow over a tube bank. The expression for the pressure drop in a tube is

$$\Delta p = \zeta \frac{1}{d} \cdot \frac{w^2}{2g} \gamma$$

(4)

Putting the free gas section equal to f, and remembering that

G=w γ f and further that $\dfrac{1}{d} = \dfrac{F}{4f}$ we obtain from Equation (3)

$$\frac{\Delta p}{\Delta t} = \zeta \frac{w^3 \gamma^2 Cp}{2g\alpha}$$

(5)

The ratio of the pressure drop in the heat transfer depends, therefore, of the velocity with which the heating surfaces are swept. If we group together those quantities which depend on the velocity,

and introduce the Reynolds number $Re = \dfrac{w\gamma d}{\mu g}$ and the Nusselt

number $Nu = \dfrac{\alpha d}{\lambda}$ we obtain the following relation between heat transfer and pressure drop:

$$\frac{\Delta p}{\Delta t / \Theta} = \frac{\mu^3 g^2 Cp}{8\gamma\lambda d^2} \cdot \zeta \cdot \frac{Re^3}{Nu}.$$

utting also:

$$\zeta \frac{Re^3}{Nu} = Z,$$

(6)

enables us to write the desired relation between the heat transfer and the energy loss in the case of longi tudinal flow through a tube, referred to 1 kg of medium flowing through the exchanger as

$$\frac{\varepsilon}{\Delta t / \Theta} = \frac{\mu^3 g^2 Cp}{8\gamma\lambda d^2} \cdot z$$

(7)

Similar expressions may be obtained for the case of a tube bank with cross flow, if ζ is taken as denoting the pressure drop coefficient per tube row, if the bank is z_1 the expression for the pressure drop becomes

$$\Delta p = \zeta \cdot z_1 \cdot \frac{w^2}{2g} \gamma$$

(4a)

where w is the velocity at the narrowest point between the tubes. If s.d denotes the pitch of the tubes across the flow, and z_q denotes the number of tubes per row also in the direction across the flow,

then with $F = \pi dl z_q.z_1$ we obtain

$$\frac{\Delta p}{\Delta t/\Theta} = \frac{w^3 \gamma^2 Cp}{8g\alpha} \cdot (s -) \cdot \frac{4}{\pi} \zeta$$

(5a)

and introducing Re and Nu gives

$$\frac{\Delta p}{\Delta t/\Theta} = \frac{w^3 g^2 Cp}{8\gamma \lambda d^2} \cdot (s-1) \cdot \frac{4}{\pi} \cdot \zeta \cdot \frac{Re^3}{Nu}$$

Let us put

$$(s-1)\frac{4}{\pi} \cdot \zeta \cdot \frac{Re^3}{Nu} = Z_q$$

(6a)

and we obtain a new form of Equation (7), for the case of cross flow:

$$\frac{E}{\Delta t/\Theta} = \frac{\mu^3 g^3 Cp}{8\gamma^2 \lambda d^2} \cdot Z_q$$

(7a)

giving the relation between the energy loss and the heat transfer for cross flow.

It is seen that the first term of the Equations (7) and (7a) contains characteristic quantities of the medium and the tube diameter. This means that for a given tube diameter the heat exchanger is fully characterized by the number Z. Now the pressure drop coefficient ζ is function of Re, while Nu is a function of Re and of the Prandtl number, if we neglect the transition zone at the inlet, which is entirely permissible with cross flow exchangers many rows deep, or with longitudinal flow exchangers with relatively long tubes. But since the Prandtl number is purely a function of characteristic quantities of the medium, and we are interested only in a comparison of heat exchangers working with the same medium, and operating within the same temperature limits, Nu depends only on Re. This makes it possible to plot Nu as a function of Z. The Nu-Z diagram, therefore, gives a clear picture of the merit of a heat exchanger surface. Tube banks of different pitch are represented by different curves in the Nu-Z diagram. The higher a curve lies, the greater may be the heat transfer loading for a given energy expenditure, or, conversely, for a given surface and heat loading, the smaller the energy expenditure. Figure 1 which is drawn for gases contains curves relating to some of the most frequently used tube arrangements. In the case of cross flow, they are based on the values derived by Grimison from the tests of Pierson and Huge.

In order to make a comparison, curves for longitudinal flow have been inserted calculated with the aid of the formula given by Jung. It is seen that only at very high rates of heat transfer such as those which are achieved by the gas velocities attained in the Velox boiler, the longitudinal arrangement become more advantageous than the cross flow one. The curves giving the values of Z which are plotted in Figure 1, in a logarithmic scale is almost straight. It is, therefore, permissible when considering segments of these curves, and without introducing any appreciable error to assume the following relation

$$\log Z = B + m \log Nu$$

or,

$$Z = BNu^{m} \tag{8}$$

where B and m are constants whose value depends on the position in the diagram of the segment under consideration.

The expression for power loss can be written using Equation (7) and transforming Equation (3):

$$L = \frac{\mu^{3}g^{2}}{8\gamma^{2}d^{3}} \cdot F \cdot B \cdot Nu^{m+1} \tag{9}$$

- Crossflow, staggered tubes, pitch 1.25 × 1.25.
- Crossflow, tubes in line, pitch 1.25 × 1.25.
- Crossflow, tubes in line, pitch 1.5 × 1.5.
- Crossflow, tubes in line, pitch 2 × 2
- Crosssflow, tubes in line, pitch 3 × 3.
- Longitudinal flow in a tube.
- Longitudinal flow between tubes, pitch 1.5 × 1 .5.
- Longitudinal tinybetween tubes, pitch 2 × 2.
- Single tube in crossflow.

The pitch is expressed as times the tubes diameter.

The diagram shows for some common tube arrangements, the relation between the heat transfer number No and characteristic number Z for the energy loss. For any given arrangement of the tubes, there is a definite value of Z for every value of Nu of heat transfer, with the help of which the pressure drop and the exchanger surface may be obtained from the Equation (7) or (7a) [3].

THE CONDITION FOR THE RIGHT EXCHANGER

The merit of an exchanger can only be judged when it is known what quantity of heat is equivalent to the mechanical energy which has to be supplied in the form of compressor or pump work to overcome the resistance of the exchanger. If the exchanger is an air preheater forming part of a steam power unit or a gas turbine then the overall efficiency of the plant or the efficiency when the exchanger is in service, determines the amount of the energy (expressed in kcal), to the heat consumption required to produce this energy [4,5].

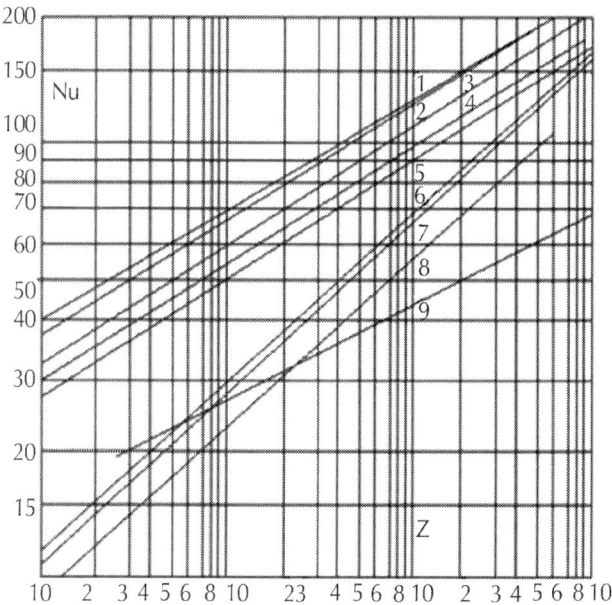

Figure 1: Relation between the energy loss and the heat transfer.

On the other hand, if it is a case of plain heat transfer for instance, in blast heaters, furnaces, etc., where the energy absorbed in overcoming the resistance of the exchanger has to be supplied in the form of power purchased from an outside supply, then the cost of this power must be balanced against the cost of production of the more or less completely transferred heat in the exchanger. If h denotes the efficiency of the plant with which the heat energy of the fuel is converted into mechanical energy, then the economic performance of the exchanger is given by

$$Useful = Q - \frac{AL}{\eta}.$$

useful heat is, therefore, equal to the difference of the heat transferred and the heat required for the production of the mechanical work absorbed.

The right heat exchanger is, therefore, the exchanger which with a given surface and with a given diameter of tubes results in a maximum amount of useful heat. The condition for this is

$$\eta dQ - AdL = 0 \tag{10}$$

We shall now seek expressions for dQ and dL in terms of Nu.

Let the suffix 1 denote the hot medium, and the suffix 2 the cold one. The meaning of the symbols is made clear by the Figure 2. We may write for the temperature variation of the hot medium

$$t_1' - t_1'' = \varepsilon_1 \cdot \left(t_1' - t_2' \right).$$

and similarly, that of the colder one is

$$t_2' - t_2'' = \varepsilon_2 \cdot \left(t_1' - t_2' \right).$$

Further, let the mean temperature difference be given by

$$\Theta = a \cdot \left(t_1' - t_2' \right).$$

The factor a depends only on e_1 and e_2, and is plotted in Figure 3 for counter flow, for cross flow and parallel flow, for the case $e_1 = e_2$. If we denote by k the overall heat transfer coefficient, the heat transferred is given by

$$Q = a \cdot k \cdot F \left(t_1' - t_2' \right) \tag{11}$$

For a small change in the rate of heat transfer we have.

$$dQ = \left(a + kF \frac{da}{dkF} \right) \cdot dkF \cdot \left(t_1' - t_2' \right).$$

Since kF may be treated as a single quantity.

We put

$$dQ = b \cdot dkF \cdot \left(t_1' - t_2' \right) \tag{12}$$

where b depends only on e_1 and e_2. From Equations (11) and (12), and from the relation $Q = \varepsilon GCp(t'_1 - t'_2)$ it is found that

$$b = \frac{a}{1 - \dfrac{\varepsilon\ da}{a\ d\varepsilon}}$$

(13)

Curves for b are given in Figure 3. In the case of heat transfer in metal exchangers the thermal resistance of the exchanger wall may be neglected without introducing any appreciable error. Hence, it is permissible to write

$$\frac{1}{kF} = \frac{1}{\alpha_1 F_1} + \frac{1}{\alpha_2 F_2}$$

(14)

Differentiating and introducing the Nusselt number in place of the quantities $d\alpha_1$ and $d\alpha_2$ gives

$$dkF = \left(\frac{kF}{\alpha_1 F_1}\right)^2 \cdot F_1 \cdot \frac{\lambda_1}{d_1} \cdot dNu_1 + \left(\frac{kF}{\alpha_2 F_2}\right)^2 \cdot F_2 \cdot \frac{\lambda_2}{d_2} \cdot dNu_2$$

(15)

The total power loss is the sum of the losses for the hot and cold mediums. We use Equation (9) and put

$$F \frac{\mu^3 g^2 B}{8\gamma^2 d^3} = P$$

(16)

and obtain for the total energy loss

$$L = P_1 Nu_1^{m_1+1} + P_2 Nu_2^{m_2+1} .$$

The change in loss with change of velocity is then given by

$$dL = P_1(m_1 + 1)_1\ Nu_1^{m_1}\ dNu_1 + P_2(m_2 + 1)\ Nu_2^{m_2}\ dNu_2$$

(17)

The change in heat quantity transferred with change of specific loading kF is given by Equation (12). Inserting Equations (12) and (17) in Equation (10) and taking into account Equation (15) gives

$$\frac{\eta b}{A}(t_1' - t_2')\left[\left(\frac{kF}{\alpha_1 F_1}\right)^2 \cdot F_1 \cdot \frac{\lambda_1}{d_1} \cdot dNu_1 \right.$$

$$+\left(\frac{kF}{\alpha_2 F_2}\right)^2 \cdot F_2 \cdot \frac{\lambda_2}{d_2} \cdot dNu_2$$

$$\left. -P_1(m_1 + 1)Nu_1^{m_1} dNu_1 - P_2(m_2 + 1)Nu_2^{m_2} dNu_2 \right] = 0 \tag{18}$$

The coefficients of dNu_1 and dNu_2 must be equal 0. This gives two new equations namely,

$$\frac{\eta b}{A}(t_1' - t_2')\left[\left(\frac{kF}{\alpha_1 F_1}\right)^2 \cdot F_1 \cdot \frac{\lambda_1}{d_1}\right] = P_1(m_1 + 1)Nu_1^{m_1} \tag{19a}$$

$$\frac{\eta b}{A}(t_1' - t_2')\left[\left(\frac{kF}{\alpha_2 F_2}\right)^2 \cdot F_2 \cdot \frac{\lambda_2}{d_2}\right] = P_2(m_2 + 1)Nu_2^{m_2} \tag{19b}$$

Dividing Equation 19 (b) by Equation 19(a), inserting for P_1 and P_2 the values given by Equation (16) and substituting Nu for a_1 and a_2 gives

$$\frac{Nu_2^{m_2+2}}{Nu_1^{m_1+2}} = \frac{(m_1 + 1) \cdot B_1 \cdot F_1^2 \mu_1^3 \gamma_2^2 \lambda_1 d_2^4}{(m_2 + 1) \cdot B_2 \cdot F_2^2 \mu_2^3 \gamma_1^2 \lambda_2 d_1^4} \tag{20}$$

This equation determines the ratio of the velocities in the right heat exchanger; it does not, however, say anything about the absolute value of the velocities. It means that heat exchangers in which this ratio of the velocities is observed have for a given surface and a given heat quantity the lowest friction loss.

If we take the roots of Equations 19(a) and 19(b), and remembering that $\dfrac{kF}{\alpha_1 F_1} + \dfrac{kF}{\alpha_2 F_2} = 1$ we obtain as the second condition for the right heat exchanger:

$$\sqrt{\frac{\eta}{A}} b(t_1' - t_2') = \sqrt{R_1 Nu_1^{m_1}} + \sqrt{R_2 Nu_2^{m_2}} \tag{21}$$

The factor R contains only constants.

$$R = (m+1)\frac{\mu^3 g^2 B}{8\gamma^2 d^2 \lambda}.$$

Two Equations (20) and (21) completely determine the most favourable heat exchanger. The resulting transcenddental equation must be solved by trial. The surfaces are found with the aid of the numbers Nu, and Nu, and since Nu is a function of the Reynolds number, the velocities w_1 and w_2 are also determinate. The dimensions of the heat exchanger are, therefore, fixed.

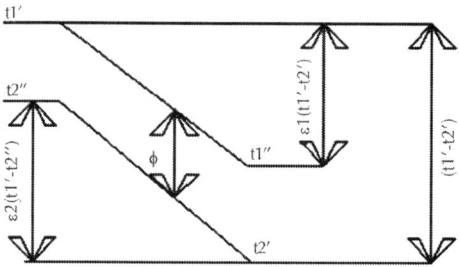

Figure 2: Diagram of temperature variation in an exchanger.

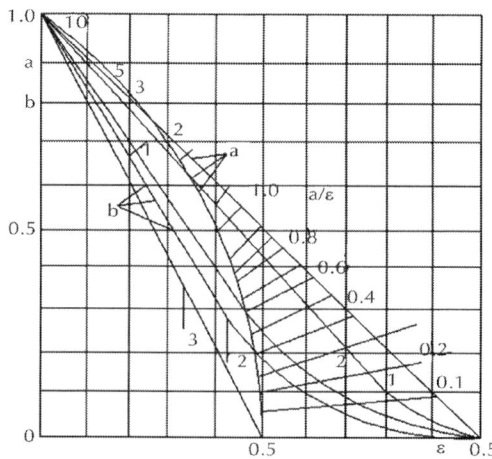

Figure 3: Characteristic numbers a and b of Equations (11) and (12) for e_1 = e_2 in the case of counterflow, crossflow and parallelflow.

THE ECONOMIC DIMENSIONING OF A HEAT EXCHANGER

It was seen in the first part that there is a function Z which serves as a criterion of the merit of tube arrangements in heat exchangers. In the second part the conditions fix the dimensions of the surface and sections of the right heat exchanger. The exchanger should, however, like every other apparatus be correctly dimensioned from the economic point of view, that is the total sum of the capital charges and of the running costs should be a minimum.

If P denotes the capital cost, n the interest and deprecition rate, then the capital chargers are

$$K_1 = nP .$$

and if L is the power absorbed, h the efficiency, h the operating hours in one year and p the price per kilowatt-hour, then the power costs are

$$K_2 = Lh\frac{1}{\eta}p .$$

and the total costs in one operational year

$$K_1 + K_2 = nP + h\frac{1}{\eta}pL .$$

there should be a minimum hence,

$$nP + h\frac{1}{\eta}pL = 0 \tag{22}$$

The capital costs will increase approximately in proportion to the exchanger surface, and for a given tube diameter, inversely proportionally to the heat transfer number, or

$$F = Nu^{-1} , P = Nu^{-1} .$$

But according to Equation (9), the energy proportional to F Nu^{m+1}; hence substituting for FNu^{-1}

$$L = CNu^m \text{ or } P \sqcap L^{\frac{1}{m}}.$$

differentiating and dividing by P

$$\frac{dP}{P} + \frac{1}{m}\frac{dL}{L} = 0 \tag{23}$$

Dividing Equation (22) by Equation (23)

$$nP + h\frac{1}{\eta} pmL = 0 \tag{24}$$

The total costs in an operational year are a minimum when the capital charges amount to m-times the power costs. Within the range of practical application, that is, for Nu = 40 to 120 the curve which is over this range can be seen upon as a straight line, for instance curve 3 (cross flow heat exchanger with a tube pitch 1.5 × 1.5) gives an exponent m = 3.84 and a constant B = 166; for curve 7 (longitudinal flow with a tube pitch 1.5 d) the figures are m = 2.67 and B = 99 500.

The starting point for this study has been the as sumption of a fixed tube diameter and tube pitch. These and the choice between staggered or straight arrangement of the tubes are determined by dirt deposit and cleaning considerations. How closely these assumptions and the results of the calculation of the right heat exchanger may be adhered to in practice depends on manufacturing conditions, but in any case the above exposition serves as a guide to show in what direction and to what extent modifications are desirable.

EXAMPLE

Designing a compact heat exchanger for heating wings continuous 20,000 kg/h of air from an inlet temperature (T_{iair}) 10°C to an out temperature (air $T_{o\,air}$) of 55°C heating fluid used water volumetric flow $\dot{V}_w$ (10 m³/h at temperature T of 95°C, the input date are shown in Table 1 [6,7].

The density and heat capacity of both fluids are obtained from tables to standard atmosphere conditions. The characteristic of tube and of the wins are shown in Table 2.

Diagram inlets from primary and secondary of the compact heat exchanger are show in Figure 4.

Unit Thermal Water Consumption

$$q_{t\,agua} = \dot{m}_{agua} cp_{agua} = \dot{V}_{agua} \rho_{agua} cp_{agua}$$

$$q_{t\,agua} = 10\,\frac{m^3}{h} \times \frac{1h}{3600s} \times 962.036\,\frac{kg}{m^3} \times 4217\,\frac{J}{kgC}$$

$$= 11269.18\,\frac{W}{0_C} = 11.269\,\frac{kW}{0_C}$$

Unit Thermal Air Consumption

$$q_{t\,aire} = \dot{m}_{aire} cp_{aire}$$

$$q_{t\,aire} = 20000\,\frac{kg}{h} \times \frac{1h}{3600s} \times 1.005\,\frac{kJ}{kgK}$$

$$= 5.577\,\frac{kW}{C}$$

$$T_{m\,aire} = \frac{55+10}{2} = 32.5°C$$

The density of air to T_m is show in Table 3.

Themal Flow

$$\dot{Q} = q_{t\,air}\left(T_{s\,air} - T_{e\,air}\right) = \dot{m}_{air}cp_{air}\left(T_{s\,air} - T_{e\,air}\right)$$

$$\dot{Q} = 5577,75\frac{W}{°C}$$

$$\left(55°C - 10°C\right) = 250.99KW = 250998,75W$$

Temperature of Exit of the Water

$$T_{s\,water} = T_{e\,water} - \frac{\dot{Q}}{q_{t\,water}}$$

$$= 95°C - \frac{250998,75W}{11269,18\,W/°C} = 73.51°C$$

Velocities are: 2.5 m/s and 3.5 m/s

$$W_{water} = 1\ \mathrm{m/s}$$

$$W_{air} = 3\ \mathrm{m/s}$$

Logarithmic Mean Temperature

$$\Delta T_{ml} = \frac{\Delta\theta_1 - \Delta\theta_2}{\ln\dfrac{\Delta\theta_1}{\Delta\theta_2}} = \frac{\left(95-55\right)-\left(73.51-10\right)}{\ln\dfrac{40}{63.51}}$$

$$\Delta T_{ml} = 50.85°C$$

Surface Airflow

$$S_p = \frac{\dot{V}_{air}}{W_{air}}$$

$$\dot{V}_{air} = \frac{W_{air}}{\rho_{air}}$$

Table gives the density of dry air at a temperature of 58.38°C

$$\rho_{air} = 1.15 \frac{kg}{m^3}$$

Volumetric Air Flow

$$\dot{V}_{air} = \frac{20000 \frac{kg}{h} \times \frac{1h}{3600s}}{1.15 \frac{kg}{m^3}} = 4.83 \frac{m^3}{s}$$

$$S_p = \frac{4.83 \frac{m^3}{s}}{3 \frac{m}{s}} = 1.61 m^2$$

$$\frac{d_{int}}{d_{out}} = \frac{11.66mm}{15.88mm} \qquad h = 40mm$$

Surface Water Flow

$$S_T = \frac{\pi}{4} d_{int}^2 N_A; \qquad W_{water} = 1 \ m/s$$

$$S_T = \frac{\dot{V}_{water}}{W_{water}} = \frac{10 \frac{m^3}{h} \times \frac{1h}{3600s}}{1 m/s} = 2.88 \times 10^{-3} \ m^2$$

number of tubes in the direction of height (h) heat exchanger

$$N_A = \frac{S_T}{A_{tube,inlet.}} = \frac{4}{\pi} \frac{S_T}{d_{inlet}^2}$$

$$= \frac{4 \times 2.78 \times 10^{-3} \, m^2}{\pi (0.01128) m^2} = 29 \, tubes$$

Height of heat exchanger

$$H = h \cdot N_A = 0.040 \, m \cdot 29 = 1.15 \, m$$

Winged tube length

$$S_p = H \cdot La \therefore La = \frac{S_p}{H} = \frac{1.61 m^2}{1.15 m} = 1.40 m$$

In Table 4 are shown calculated values of Na, H and L to the compact heat exchanger.

Wing thickness, this value is assumed by the types of materials available in the industry e=0.3mm.

Calculation of Reynolds Number (Re), Nusselt (Nu) and the Convective Coefficient (hi) of primary fluid Whereas, dh = di; and the Prandtl Number (Prp), dynamic viscosity (mp) and conductivity of the primary fluid are determined from tables to the average temperature of primary fluid (Tmp). These values are shown in Table 5.

$$Rep = \frac{Wp \cdot dh}{\mu p}$$

$$Nu = 0.023 \cdot Rep^{0.8} \cdot Prp^{0.3}$$

$$hi = \frac{Nu \cdot kp}{dh}$$

The values of Reynolds number, Nusselt number and convection coefficient inside of primary fluid are shown in Table 6.

Calculation of the mass velocity and maximum mass velocity of the secondary fluid.

$$s = \frac{h}{de}$$

$$Wms = Ws \cdot \rho s$$

$$Wms' = Wms \left(\frac{s}{s-1} \right)$$

The values of mass velocity, maximum mass velocity and relationship s are shown in Table 7.

Calculation of hydraulic diameter (dh¢), Reynolds (Re) and convection coefficient (I) as a function the secondary fluid pitch between wing (p) Tables obtained forthe Prandtl number (Pr), the dynamic viscosity (ms) and thermal conductivity (ks) of secondary fluid at film temperature (Tmp). These values are shown in Table 8.

$$dh' = 2p \left(1 - \frac{\pi \cdot de^2}{4h \cdot l} - \frac{e}{p} \right)$$

$$Re = \frac{Wms' \cdot dh'}{\mu s}$$

$$Nu = 0.26 \cdot Re^{0.6} \cdot Prp^{1/3}$$

$$he = \frac{Nu \cdot ks}{dh'}$$

Table 9. show the values of m, dh', Re, Nu and he of the secondary fluid.

Calculation of equivalent diameter (dea) and efficiencies of wings. (Table 10)

In Table 11 show characteristics of wing, Aluminum considering.

$$dea = \frac{2 \cdot h \cdot l}{h + l}$$

$$\mu = \frac{dea}{de}$$

$$\beta = \sqrt{\frac{2 \cdot he}{ks \cdot e}}$$

$$\left[de \cdot \beta \cdot \frac{\mu - 1}{2} \right]_{,}$$

Relationship for the efficiencies.

Calculation of the ratio number of wings with borders with the total number of wings.

Considering that N = Na x Np where Na is the number of tubes in direction of height and Np is the number of tubes in the direction of secondary flow.

Number of wings without borders

$$_* N0 = Na \cdot Np - 2(Na + Np) + 4$$

Number of wings with a border

$$_* N1 = 2(Na + Np) - 8$$

Number of wings with two borders

$$_* N2 = 4$$

*Note: This means that the equations are modified as a function on Np. (Table 12)

$$\psi 0 = \frac{N0}{N} \quad \psi 1 = \frac{N1}{N} \quad \psi 2 = \frac{N2}{N}$$

Calculation of correction coefficients for the wings. (Table 13)

$$C_0 = 1$$

$$C_1 = 1 + 0.5 \left[\frac{h_e}{k_{al}} \left(\frac{l}{h} + \frac{h}{l} \right) \right]^{1/4}$$

$$C_2 = 1 + \left[\frac{h_e}{k_{al}} \left(\frac{l+h}{h} \right) \right]^{1/3}$$

Calculation of the overall efficiency of the equivalent circular wing. (Table 14)

$$\eta g = \eta f \left[\frac{\psi_0}{C_0} + \frac{\psi_1}{C_1} + \frac{\psi_2}{C_2} \right]$$

Calculation the overall surface heat Exchange where, $X = Np.l$. (Table 15)

$$Se = (La \cdot X \cdot H) \left(\left(\frac{\pi \cdot de}{h \cdot l} \right) + \frac{2}{p} - \frac{\pi \cdot de^2}{2 \cdot p \cdot h \cdot l} \right)$$

Calculating the overall coefficient of heat transfer function and pitch Np

$$U = \frac{1}{\frac{Se}{Si} \left(\frac{1}{hi} + \frac{et}{kt} \right) + \frac{1}{he} \left[1 - \frac{Ss}{Se} (1 - \eta g) \right]}$$

where,

$$Si = \pi \cdot dik \cdot La \cdot Na \cdot Np$$

$$Ss = 2 \cdot \frac{La}{p} \left(X \cdot H - Np \cdot Na \cdot \left[\frac{\pi \cdot de^2}{4} \right] \right)$$

*kt = conductivity of tube, e_t = Thickness of the wall of tube.

**The value $\frac{et}{kt}$ is neglected as being small compared to $\frac{1}{hi}$.
(Table 16)

Calculation Heat flow based on Np and pitch (p)

$$Q = U \cdot A \cdot (\Delta T)$$

where A = Se = Overall surface heat exchange DT = Tml = Difference logarithmic mean temperature. (Table 17)

Calculation U.Np.St$_1$ for comparison of the relationship (Q / °C) as a function of Np and the pitch between wings.

$$Q = U \cdot Np \cdot St_1 \cdot (\Delta\theta_{ml})$$

$$\therefore U \cdot Np \cdot St_1 = \frac{Q}{\Delta\theta_{ml}} = \frac{250998,75}{50.85} = 4935,724 \ W/°C$$

Where:

St = Overall surface heat exchange with respect to pitch

Np = Pitch number of tubes in the direction of seconddary flow. (Table 18)

Selection based on the results obtained in the above table

U.Np.St$_1$ $= \dfrac{Q}{\Delta\theta_{ml}}$ can be compared to select the exchange with their respective sizes, data are compared

$$4935,724059 \approx 5300,392$$

This variation is important for any factor that our team needs to transfer more heat.

The selection of compact heat exchanger required to have 6 columns of tubes (Np = 6) in the direction of flow and pitch inlet wings of 8 mm, so X will be worth

$$X = 1 x Np = (0.008)(6) = 0.18m$$

The dimensions of the compact heat exchanger will (L) x(H) x(X), (1.40 × 1.15 × 0.18) m.

With piping HWG 22 de = 12.7 mm y di = 11.28 mm.

The thick aluminum wings e = 0.3 mm and K = 203.52 W/m C. In Table 19 are shown characteristics of the exchanger.

Table 1: Input data of the compact heat exchanger

Input data		
T_{inlet} of Secondary fluid (Tfi)	10.00	°C
T_{out} of Secondary fluid (Tfo)	55.00	°C
T_{inlet} of Primary fluid (Tci)	95.00	°C
velocity primary fluid (Wp)	1.00	m/s
velocity secondary fluid (Ws)	3.00	m/s
Mass flow of primary fluid (mp)	2.77	Kg/s
Mass flow of secondary fluid (ms)	5.55	Kg/s
heat capacity primary fluid (Cpp)	4217	J/kg °C
heat capacity secondary fluid (Cps)	1005.00	J/kg °C
density primary fluid (ρp)	962.04	Kg/m³
density secondary fluid (ρs)	1.15	Kg/m³

Table 2: Characteristic of tube and of the wing

Characteristic of tube		
Outside diameter of tube (do)	0.012700	m
Inside diameter of tube (di)	0.011280	m
Wheelbase of the tubes in the direction of the tubes (l)	0.080	m
Wheelbase of the tubes in the direction of height (h)	0.040	m
Area inside of the tube (At-int)	9.993E - 505	M²
Characteristics of the wing		
Thickness of the wing (e)	3.00E - 04	m

Thermal conductivity (Aluminum , k)	203.52	W/m °C

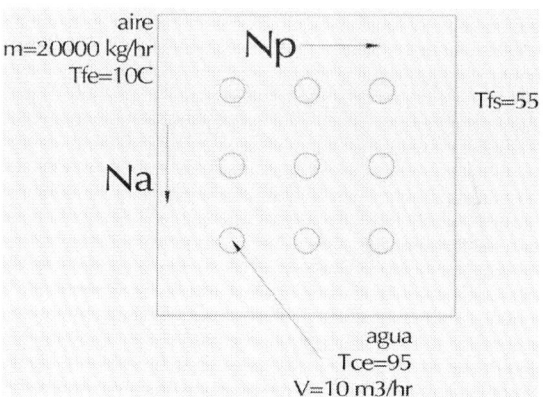

Figure 4: Diagram inlets from primary and secondary fluids.

Table 3: Density of air to T_m

$T(°C)$	$\rho(\dfrac{kg}{m^3})$
30	1.165
32.5	$\rho_{air} = 1.15(\dfrac{kg}{m^3})$
40	1.128

Table 4: Values of Na, H and L to the compact heat exchanger

Number of tubes in direction of height (Na)	29	Tubos
Exchanger height (H)	1.15	m
Overall longitude of the tube (L)	1.40	m

Table 5: Values of Prp, v and kp

Prandtl number of primary fluid (Prp)	2.11	
Viscosity of primary fluid (v)	3.50E-07	m^2/s
Conductivity of primary fluid (kp)	0.68	w/m°C

Table 6: Values of Rep, Nup and hi to the compact heat exchanger

Reynolds number of primary fluid (Rep)	32228.57	
Nusselt number of fluid primary (Nup) 1	116.31	
Convection coefficient inside (hi)	6972.23	w/ m^2-K

Table 7: Values of s, Wms and Wms of the secondary fluid

Relationship between h and the outer diameter (s)	2.36	
Mass velocity of secondary fluid (Wms)	3.45	Kg/s-m^2
Maximus velocity of secondary fluid (Wms')	5.98	Kg/s-m^2

Table 8: Values of Tp, Pr, rs and ks of secondary fluid

| Film Temperature (Tp) | 58.33 | °C |
| Prantl Number of secondary fluid (Prs) | 0.6963 | |

| Dynamic Viscosity of secondary fluid (ρs) | 0.0781 | Kg/m-s |
| Conductividad del flujo se-cundario (ks) | 0.0287 | w/m $^{\circ}$C |

Table 9: Values of m, dh¢, Re, Nu and he of the secondary fluid

Pitch (m)	Hydraulic Di-ameter (dh¢)	Reynolds Number sec-ondary (Re)	Nusselt Num-ber second-ary (Nu)	Convective Coefficient secondary (he)
0.002	3.08E - 03	718.4305	11.923624	110.9873
0.003	4.92E - 03	1147.548	15.792099	92.027767
0.004	6.77E - 03	1576.665	19.108255	81.04597
0.005	8.61E - 03	2005.782	22.077386	73.606079
0.006	1.04E - 02	2434.899	24.800817	68.113749
0.007	1.23E - 02	2864.017	27.337746	63.831783
0.008	1.41E - 02	3293.134	29.726447	60.364755

Table 10: Values of equivalent diameter and efficiencies of wings

Pitch (n)	Convective Coefficient secondary (he)	Equivalence Factor wings (TO	den** = (de *ᵢfi) *(fi1-1/2)	Efficiency obtained from the graphic of
0.002	110.9873	60.295115	0.823028315	0.9
0.003	92.027767	54.904119	0.749441229	0.92
0.004	81.04597	51.524193	0.703305238	0.93
0.005	73.606079	49.102356	0.670247159	0.95

0.006	68.113749	47.234887	0.644756207	0.96
0.007	63.831783	45.726081	0.624161001	0.97
0.008	60.364755	44.466936	0.606973681	0.98

Table 11: Characteristics of Aluminum

| Thermal conductivity of aluminum (k) | 203.525 | w/mC |
| Wing thickness (e) | 3.00E-04 | m |

Table 12: Relations with number of wings on the total number of wings

Relations with number of wings on the total number of wings						
ψ	(Np) = 1	(Np) = 2	(Np) = 3	(Np) = 4	(Np) = 5	(Np) = 6
$\tilde{\psi}_0$	0	0	0.3101952	0.4652928	0.5583514	0.620390394
ψ_1	0	0.930586	0.6435285	0.5	0.4138829	0.35647147
ψ_2	0.930585591	0.069414	0.0462763	0.0347072	0.0277658	0.023138136

Table 13: Correction factors for each type of wing

Correction factors for each type of wing				
Pitch (m)	h_e	C_0	C_1	C_2
0.002	110.987	1	1.510965	2.029347
0.003	92.0278	1	1.487588	1.967038
0.004	81.046	1	1.472341	1.926932
0.005	73.6061	1	1.461107	1.897653
0.006	68.1137	1	1.452253	1.874747
0.007	63.8318	1	1.444972	1.856018
0.008	60.3648	1	1.438802	1.840231

Table 14: Overall efficiency (hg) as a function of Np and pitch (p)

	Overall efficiency (ηg) as a function of Np and pitch (p).					
ηf	(Np) = 1	(Np) = 2	(Np) = 3	(Np) = 4	(Np) = 5	(Np) = 6
0.9	0.412708	0.585084	0.683014	0.731979	0.761358	0.780944
0.92	0.435243	0.607987	0.705014	0.753528	0.782636	0.802041
0.93	0.449131	0.621303	0.717299	0.765297	0.794095	0.813294
0.95	0.465868	0.639809	0.736269	0.784499	0.813437	0.832729
0.96	0.476524	0.650701	0.746883	0.794974	0.823829	0.843066
0.97	0.486346	0.660974	0.757071	0.80512	0.833949	0.853168
0.98	0.495576	0.670808	0.766957	0.815031	0.843876	0.863105

Table 15: Total surface heat exchange

	Overall efficiency (ηg) as a function of Np and pitch (p).					
P(m)	(Np) = 1	(Np) = 2	(Np) = 3	(Np) = 4	(Np) = 5	(Np) = 6
0.002	60,85782	121,7156	182,5734	243,4313	304,2891	304,2891
0.003	41,10674	82,21349	123,3202	164,427	205,5337	205,5337
0.004	31,23121	62,46242	93,69363	124,9248	156,156	156,156
0.005	25,30589	50,61178	75,91766	101,2236	126,5294	126,5294
0.006	21,35567	42,71135	64,06702	85,42269	106,7784	106,7784
0.007	18,53409	37,06818	55,60228	74,13637	92,67046	92,67046
0.008	16,41791	32,83581	49,25372	65,67162	82,08953	82,08953

Table 16: Overall rate of heat transfer (W/m² C)

	OVERALL RATE OF HEAT TRANSFER (W/m² C)					
P(m)	(Np) = 1	(Np) = 2	(Np) = 3	(Np) = 4	(Np) = 5	(Np) = 6
0.002	100.1748	86.99647	80.94672	78.22676	76.68079	84.68974
0.003	109.8194	91.66171	83.87258	80.4542	78.53373	84.50856
0.004	110.6942	90.50486	82.55818	78.52662	76.50163	81.06726
0.005	107.1119	86.58511	65.14468	74.68027	72.68163	76.02262
0.006	102.8529	82.72923	62.38494	71.19159	69.25976	71.86373
0.007	98.2215	78.86459	72.69445	67.83032	65.9839	68.03117
0.008	93.64498	75.20035	69.27388	64.71031	62.95396	64.56843

Table 17: Values of heat flow (Q)

Heat flow (Q)						
P(m)	(Np) = 1	(Np) = 2	(Np) = 3	(Np) = 4	(Np) = 5	(Np) = 6
0.002	310024.2	538478.9	751549.4	968394.7	1186571	1310502
0.003	229568.8	383223.1	525987	672732.6	820842.8	883292.4
0.004	175806.2	287482.5	393360.6	498868.8	607505.5	643761.5
0.005	137841.4	222851.5	251502.6	384421.9	467667.2	489164.7
0.006	111699.4	179689.6	203252.1	309259.2	376084.1	390223.8
0.007	92576.03	148663.4	205548.6	255726.6	310956.7	320604.8
0.008	78184.91	125570.7	173511.9	216108.5	262803.7	269543.4

Table 18: Values of $U \cdot Np \cdot St_1$ (W/°C)

$U \cdot Np \cdot St_1$ (W/°C)						
P(m)	(Np) = 1	(Np) = 2	(Np) = 3	(Np) = 4	(Np) = 5	(Np) = 6
0.002	6096,42	10588,83	14778,72	19042,84	23333,13	25770,16
0.003	4514,317	7535,829	10343,19	13228,84	16141,33	17369,36
0.004	3457,113	5653,152	7735,176	9809,925	11946,19	12659,14
0.005	2710,561	4382,226	4945,632	7559,402	9196,366	9619,099
0.006	2196,494	3533,477	3996,817	6081,377	7395,444	7673,492
0.007	1820,446	2923,367	4041,977	5028,693	6114,758	6304,48
0.008	1537,454	2469,264	3411,996	4249,631	5167,861	5300,392

CONCLUSIONS

This paper identifies the advantages of having the appropriate exchanger with working conditions, environmental conditions and economic aspects, it is also necessary to mention the following regarding the general utility of this work.

- In addition to the thermal design, mechanical design of heat exchangers is also a part of it. The mechanical design is done under the ASME Section VIII, which is entitled "Pressure Vessel Design"

• Although the subject of this work is the design of heat exchangers, which, as noted was achieved successfully, the utility of it is wider and there are several methods for the design of heat exchangers.

Table 19: Characteristics of the exchange

CHARACTERISTICS OF THE EXCHANGE		
de	Outside diameter of the tubes	12.7m
di	Internal diameter of the tubes	11.28m
La	Wings tube length	1.40m
L	Overall length of the tube.	1.40m
N_a	Number of vertical tubes in the exchanger	29 tubos
N_p	Number of tubes in the direction of secondary flow.	6 tubos
N	Overall number of tubes (N= Na Np).	174 tubos
h	Wheelbase of the tubes in the vertical direction.	0.04m
l	Wheelbase of the tubes in the direction of secondary flow	0.08m
n	Number of wings (n = La/p)	175 wings
p	Wings pitch (from center to center of wings)	0.008 m
e	Wing thickness.	0.3 mm
H	Height of heat exchanger (continuous wing)	1.15 m
X	Heat exchanger width (within the meaning of secondary flow)	0.18 m

• While this paper addresses just one example of a heat exchanger, this could vary, the same results in the thermal design, these variations are affected primarily for economic reasons or space.

REFERENCES

1. D. Q. Kern, "Procesos de Transferencia de Calor," 31th Reimpresión, CECSA, México, 1999.

2. F. P. Incropera and D. P. Dewitt, "Fundamentals of Heat and Mass Transfer," 5 th Edition, Wiley, New York, 2002.

3. M. Serna and A. Jimenez, "A Compact Formulation of Bell-Delaware Method for Heat Exchanger Desing and Optimation," Chemical Engineering Research and Desing, Vol. 83, (A5), 2005, pp. 539-550.

4. R. K. Shah, P. Dusan and D. P. Sekulic, "Fundamentals of Heat Exchanger Design," John Wiley & Sons, 2003.

5. Z. H. Ayub, "A New Chart Method for Evaluating Single-Phase Shell Side Heat Transfer Coefficient in a Single Segmental Shell and Tube Heat Exchanger," Applied Thermal Engineering, Vol. 25, 2005, pp. 2412-2420.

6. Y. A. Kara and Ö. Güraras, "A Computer Program for shelland-tube Heat Exchanger," Applied Thermal Engineering, Vol. 24, 2004, pp 1797-1805.

7. R. Castillo, "Diseño Computacional de Intercambiadores de Calor de Coraza y Tubos por el Método Delaware," Tesis de Maestría, Instituto Politécnico Nacional, México, 1999.

Delaware Method Improvement for the Shell and Tubes Heat Exchanger Design

Miguel Toledo-Velázquez[1], Pedro Quinto-Diez[1],
Juan C. Alzelmetti-Zaragoza[2], Sergio R. Galvan[3],
Juan Abugaber-Francis[1], and Arturo Reyes-León[1]

[1]Applied Thermal and Hydraulic Engineering Laboratory SEPI-ESIME-IPN Professional Unit "Adolfo Lopez Mateos", México DF, México

[2]Faculty of Mechanical Electrical Engineering, Universidad Veracruzana, Veracruz, México

[3]Faculty of Mechanical Engineering, Universidad Michoacana de Sán Nicolas de Hidalgo, Edifice H, University City, Santiago Tapia 403, Col. Centro, Morelia, Michoacán, México

ABSTRACT

In this paper the Delaware Method published in 1963 is analized and upgraded with using correction factors which take into account the undesirable currents of the mean flow. However, this method presents graphically these correction factors which imply an impediment to fulfill the software calculations. Thus, the equations corresponding to the correction factor equations and a Fortran 77 numerical program were established. This system is given to explore different design alternatives in order to find the optimal solution to each proposed problem. The results of this work was a simple software that can perform calculations with the introduction of parameters depending only on the geometry of the heat exchanger, i.e., geometry, temperature and fluid characteristics eliminating the human errors and increasing the calculations speed and accuracy.

INTRODUCTION

The method is established in the heat transfer analysis and in the pressure losses of the fluid which flows through the shell side [1] [2]

In order to complete the heat exchanger analysis, it is essential to consider the different currents generated by the shell side flow, as is shown in Figure 1. This figure has been modified by Palen and Taborek [3] [4] in regard to the original version proposed by Tinker [5]. Here five different streams are identified side of the shell. The current B is known as the mean mixed flow and flows through the mixed flow section and the window section of the heat exchanger. This is the ideal current flowing in the shell side of a heat exchanger.

Besides of flow B, there are four more flows which are present due to the free spaces and cavities between the shell and the deflectors and they provoke a modification of the current B performance.

The different leakages and recirculation currents influence the heat flux transfer in two ways:

- Reduce the current B resulting in a drop of the heat transfer global coefficient.
- Modify the shell-side temperature distribution.

The Delaware method considers these effects as correction factors in the heat transfer coefficient and the loss pressure calculations.

The basic equations of the heat exchangers thermal design are:

$$Q = U_{dc} A F T_{mlc} \tag{1}$$

$$U_{dc} = \cfrac{1}{\cfrac{1}{h_{io}} + \cfrac{1}{h_{cc}} + R_d + \cfrac{e_t}{k}} \tag{2}$$

In the Equation (2), h_{cc} is the fluid convection coefficient flowing in the shell and it is obtained by Equation (3). In this equation the correction factors are included, the correction factors J_i, J_c, J_l, J_b, J_r, J_r^* and J_s, which consider the different currents shown in Figure 1.

$$h_{cc} = J_i C p_c \left(\frac{W_c}{S_m} \right) \left(\frac{k_c}{C p_c \mu_c} \right)^{2/3} \left(\frac{\mu_c}{\mu_{wc}} \right)^{0.14} J_c J_l J_b J_r J_s \tag{3}$$

For the hydraulic design is required to determine the fluid pressure loss through the shell side and it must be included the correction factors f_i, R_b, R_l and R_s. The corresponding calculation is made using the next equation:

$$\Delta P_c = \left\{ (N_b - 1) \left[\frac{4 f_i W_c^2 N_c}{2 g \rho_c S_m^2} \left(\frac{\mu_{wc}}{\mu_c} \right)^{0.14} \right] R_b R_l + 2 \Delta P_{b,i} \left(1 + \frac{N_{cw}}{N_c} \right) R_b R_s + N_b \Delta P_{w,i} R_l \right\} \tag{4}$$

It is observed that the principal problem using the Delaware Method is the determination of the eleven correction factors of which nine are obtained graphically. In the next section, the nine correction factors will be presented analytically. The original version only presents analytically the J_s and R_s correction factors.

SHELL SIDE CORRECTION FACTOR EQUATIONS

According to the Delaware method, after the geometrical parameters computations, the heat transfer and loss pressure are estimated considering the correction factors. This special case will be shown in the next section by using the appropriate Equations [6] -[8] .

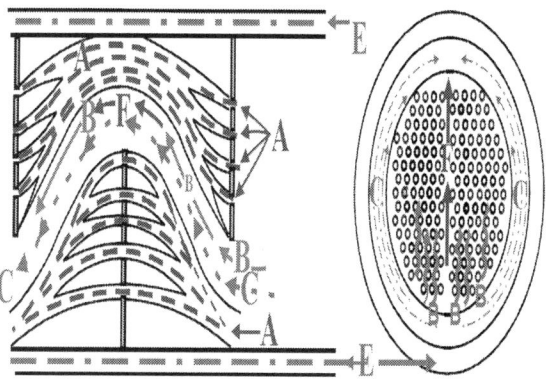

Figure 1: Ideal design of the currents flowing in the shell side. A. Currents flowing through the free spaces between tubes and baffles; B. Mean current flow; C. Recirculation current; D. Leakage current between shells and deflectors; E. Recirculation current in the past partition.

Correction Factor J_i

This factor depends on the Reynolds Number Re_c and the tube arrangement as is established in the next equation:

$$J_i = \exp\left[A + B\ln\left(Re_c\right) + C\ln\left(Re_c\right)^2 + D\ln\left(Re_c\right)^3 + E\ln\left(Re_c\right)^4 + F\ln\left(Re_c\right)^5 \right] a \tag{5}$$

This equation is applied to the different arrangement tube types: triangular, square, and rhombic. For each case, the constant are presented in Table 1. It was determined 4% as that the maximum error between the equations and the graphic presentation.

Correction Factor J_c by the Baffle Configuration Effect

The deflector geometry and especially the window cross-section shape provoke currents where the effects are considered in the correction factor J_c. This factor is determined in a mathematical way as following:

$$J_c = A + BF_c + CF_c^2 + DF_c^3 + EF_c^4 \tag{6}$$

The values of each constant are presented in Table 2. Comparing the graphics obtained by Equation (6) and the original one, the maximum error established is 5%.

Correction Factor by Baffle Leakages Effect, J_l

The fluid that does not exchange heat owing to the leakage through the gap formed by the tubes and the baffles as well as through the shell and baffle is considered in the correction factor J_l. This factor is obtained using the Equation (7) where:

$$S_1 = S_{sb}/(S_{sb} + S_{tb}) \quad \text{and} \quad S_2 = (S_{sb} + S_{tb})/S_m$$

The constants required by this equation are available in Table 3. Comparing the original graph against that one getting by Equation (7), the maximum error is 2%.

$$J_l = \left[A + B(S_1) + C(S_1)^2 + D(S_1)^3 \right] + \left[E + F(S_1) + G(S_1)^2 + H(S_1)^3 \right](S_2)$$

$$+ \left[I + J(S_1) + K(S_1)^2 + L(S_1)^3 \right](S_2)^2 + \left[M + N(S_1) + O(S_1)^2 + P(S_1)^3 \right](S_2)^3 \tag{7}$$

Table 1: Constants for the equation for the correction factor J_i

Arrangement	A	B	C	D	E	F
Triangular	0.627615	−0.69064	−0.0507472	0.0141049	−0.000937714	1.7683×10^{-5}
Square	0.374177	−0.671577	−0.0784051	0.02507191	−0.00224983	0.0000673254
Rhombic	−0.273166	−0.472896	−0.109701	0.023299	−0.00145983	0.0000242675

Table 2: Constants for the equation for the correction factor J_c

For values of F_c between	A	B	C	D	E
0 and 0.9	0.533574545	0.69059596	0.290909091	−0.295959596	0
0.9 and 1	−27.84837787	152.5274893	−301.9699773	265.12743360	−86.76640715

Table 3: Constants for the equation for the correction factor J_l

Values of S_2	A	B	C	D	E	F	G	H
0 to 0.1	1	0	0	0	-2.5903333	-4.8677761	8.773331	-7.0222218
0.1 to 0.7	0.90063003	-0.0389402	0.1363475	0.05531749	-0.4299145	-1.0949101	1.4689544	0.7341744
Values of S_2	I	J	K	L	M	N	O	P
0 to 0.1	15.8742857	50.8190476	-133.48571	106.666666	-39.6381	-168.31699	486.093325	-391.10933
0.1 to 0.7	-0.0567309	1.72685112	-4.2124254	2.4290305	0.04115536	1.3220234	3.32880672	-1.99187

Correction Factor by Recirculation Flow Effect, J_b

The recirculation currents do not exchange heat through the tubes and this issue is considered by the correction factor J_b. This factor is computed by Equation (8) which is based on F_{sbp}. The constant of the Equation $N_2 = N_{ss}/N_c$ are defined in Table 4 in terms of Re_c.

$$J_b = \left[A + B(N_2) + C(N_2)^2 + D(N_2)^3 + E(N_2)^4\right] + \left[F + G(N_2) + H(N_2)^2 + I(N_2)^3 + J(N_2)^4\right]F_{sbp}$$
$$+ \left[K + L(N_2) + M(N_2)^2 + N(N_2)^3 + \tilde{N}(N_2)^4\right]F_{sbp}^2 + \left[O + P(N_2) + Q(N_2)^2 + R(N_2)^3 + S(N_2)^4\right]F_{sbp}^3$$
$$+ \left[T + U(N_2) + V(N_2)^2 + W(N_2)^3 + X(N_2)^4\right]F_{sbp}^4 \tag{8}$$

A maximum error of 1.3% is result of comparing the original graph against that one obtained by Equation (8).

Correction Factor by Adverse Temperature Gradient J_r

This factor has the value of 1 if $Re_c \geq 100$. When Re_c fluctuation is between 0 and 100, the next criterion is used:

1. If $Re_c \leq 20$, J_r will acquire the same value as J_r^* using the Equation (9), and knowing that N_b and $N_1 = N_c + N_{cw}$ are constants which are shown in Table 5.

$$J_r^* = \left[A + B(N_1) + C(N_1)^2 + D(N_1)^3 + E(N_1)^4\right] + \left[F + G(N_1) + H(N_1)^2 + I(N_1)^3 + J(N_1)^4\right]N_b$$
$$+ \left[K + L(N_1) + M(N_1)^2 + N(N_1)^3 + \tilde{N}(N_1)^4\right]N_b^2 + \left[O + P(N_1) + Q(N_1)^2 + R(N_1)^3 + S(N_1)^4\right]N_b^3$$
$$+ \left[T + U(N_1) + V(N_1)^2 + W(N_1)^3 + X(N_1)^4\right]N_b^4 \tag{9}$$

A maximum error of 1.9% is result of comparing the original graph against that one obtained by Equation (8).

2. If $20 \leq Re_c \leq 100$, J_r is computed by Equation (10) which depends on J_r^*, Re_c and Table 6 constants.

Table 4: Constants for the equation for the correction factor J_b

For	A	B	C	D	E	F	G	H	I
$Re_c > 100$	0.99939474	0.01085613	-0.0171267	-0.1003071	0.19198207	-1.2394301	13.4729632	-85.593388	252.064457
$Re_c < 100$	0.9991889	0.02071706	-0.185339	0.59250929	-0.5964414	-1.3564474	16.400326	-116.51387	362.367998
For	J	K	L	M	N	ñ	O	P	Q
$Re_c > 100$	-249.7121	0.74529366	-9.5329177	82.1849019	-321.67178	378.959635	-0.3575412	3.12005087	-71.95138
$Re_c < 100$	-368.18466	1.18264455	-19.614535	140.62959	-451.59784	478.689065	-1.1031648	16.9758369	-113.6298
For	R	S	T	U	V	W	X		
$Re_c > 100$	422.592705	-576.65103	0.20001675	-3.866875	70.1962129	-363.1866875	473.343386		
$Re_c < 100$	396.869404	-457.40996	0.59140536	-8.7366322	56.8613935	-210.10305	253.214998		

Table 5: Constants for the equation for the correction factor J_r^*

A	B	C	D	E	F	G	H	I
1.11286866	−0.0444079	0.0019872	-4.324×10^{-5}	3.4787×10^{-7}	−0.0289065	2.241×10^{-4}	1.5142×10^{-6}	9.5563×10^{-8}
J	K	L	M	N	Ñ	O	P	Q
-3.11×10^{-9}	0.0008806	2.8301×10^{-5}	-2.088×10^{-6}	3.4527×10^{-8}	-9.296×10^{-11}	-1.363×10^{-5}	-9.749×10^{-7}	5.9493×10^{-8}
R	S	T	U	V	W	X		
-8.947×10^{-10}	1.618×10^{-12}	8.3072×10^{-8}	8.7699×10^{-9}	-4.727×10^{-10}	5.7282×10^{-12}	8.9289×10^{-15}		

Table 6: Constants for the equation for correction factor J_r

A	B	C	D	E	F	G	H
−0.2477376	0.0129611	-1.127×10^{-5}	6.5427×10^{-8}	1.26556042	−0.0129666	2.5357×10^{-6}	1.5481×10^{-8}
I	J	K	L	M	N	O	P
−0.0481867	2.2364×10^{-4}	1.6928×10^{-5}	-1.708×10^{-7}	0.03112154	-2.758×10^{-4}	-6.979×10^{-6}	8.3062×10^{-8}

$$J_r = \left(A + B\,\mathrm{Re}_c + C\,\mathrm{Re}_c^2 + D\,\mathrm{Re}_c^3\right) + \left(E + F\,\mathrm{Re}_c + G\,\mathrm{Re}_c^2 + H\,\mathrm{Re}_c^3\right)J_r^*$$
$$+\left(1 + J\,\mathrm{Re}_c + K\,\mathrm{Re}_c^2 + L\,\mathrm{Re}_c^3\right)J_r^{*2} + \left(M + N\,\mathrm{Re}_c + O\,\mathrm{Re}_c^2 + P\,\mathrm{Re}_c^3\right)J_r^{*3} \quad (10)$$

In the previous equation, J_r^* is calculated by Equation (9). Comparing both graphics, the original and that one obtained by Equation (10), it is observed as maximum error 3.8%.

Correction Factor by Uneven Baffle Spacing at the Inlet and/or Outlet, J_s

This factor has an effect when exist a different baffle distribution at the inlet and/or outlet and along the tube bundle and it is computed by Equation (11):

$$J_s = \frac{\left(N_b - 1\right) + \left(l_{s,I}^*\right)^{1-n''} + \left(l_{s,o}^*\right)^{1-n''}}{\left(N_b - 1\right) + l_{s,I}^* + l_{s,o}^*} \quad (11)$$

where $n'' = 0.6$ for turbulent flow and $(\mathrm{Re}_c > 100)$.

$n'' = 1/3$ for laminar flow and $(\mathrm{Re}_c < 100)$.

Correction Factor by Friction of an Ideal Bank Tubes f_i

The correction factor by friction in a triangular and rotated square set is determined in function of Re_c and the constants presented in Table 7 by the Equation (12).

$$f_i = \exp\left\{A + B\ln\left(\mathrm{Re}_c\right) + C\ln\left(\mathrm{Re}_c\right)^2 + D\ln\left(\mathrm{Re}_c\right)^3 + E\ln\left(\mathrm{Re}_c\right)^4 + F\ln\left(\mathrm{Re}_c\right)^5\right.$$
$$\left. + G\ln\left(\mathrm{Re}_c\right)^6 + H\ln\left(\mathrm{Re}_c\right)^7 + I\ln\left(\mathrm{Re}_c\right)^8 + J\ln\left(\mathrm{Re}_c\right)^9\right\} \quad (12)$$

A maximum error of 4.9% is result of comparing the original graph to that one obtained by Equation (12).

Pressure Loss Correction Factor by Tube-Baffle Leakage R$_l$

The introduction of this correction factor is due to the leakage of the tube bundle and it is computed by Equation (13) which is in terms of $S_1 = S_{sb} / (S_{sb} + S_{tb})$ and $S_2 = (S_{sb} + S_{tb}) / S_m$

$$R_l = \left[A + B(S_1) + C(S_1)^2 + D(S_1)^3\right] + \left[E + F(S_1) + G(S_1)^2 + H(S_1)^3\right]S_2$$
$$+ \left[I + J(S_1) + K(S_1)^2 + L(S_1)^3\right](S_2)^2 + \left[M + N(S_1) + O(S_1)^2 + P(S_1)^3\right](S_2)^3 \quad (13)$$

The constants required by the equations are established in Table 8. Comparing the original graph and that one obtained by Equation (13) is observed a maximum error of 4.5%.

Table 7: Constants for the equation for the correction factor f_i

Type arrangement, outside diameter and pitch between tubes			A	B	C	D	E
Arrangement	**Diameter**	**Pitch**					
Triangular	19.05 mm	23.8125 mm	4.15076	-0.675323	-0.254615	0.0590471	-0.00431448
Triangular	25.4 mm	31.75 mm	4.15076	-0.675323	-0.254615	0.0590471	-0.00431448
Rhombic	19.05 mm	25.4 mm	3.69311	-1.18662	0.281179	-0.136488	0.0280708
Triangular	19.05 mm	25.4 mm	3.85004	-0.609235	-0.278939	0.0630913	-0.00452036
Rhombic	25.4 mm	31.75 mm	3.97656	-0.796179	-0.1565	0.0319655	-0.00143937
Square	19.05 mm	25.4 mm	3.76203	-0.9323	-0.0827537	0.0678788	-0.0281861
Square	25.4 mm	31.75 mm	3.99352	0.768721	-0.32634	0.176985	-0.0326325
Type arrangement, outside diameter and pitch between tubes			F	G	H	I	J
Arrangement	**Diameter**	**Pitch**					
Triangular	19.05 mm	23.8125 mm	0.000102933	0	0	0	0
Triangular	25.4 mm	31.75 mm	0.000102933	0	0	0	0
Rhombic	19.05 mm	25.4 mm	-0.00237654	0.0000713989	0	0	0
Triangular	19.05 mm	25.4 mm	0.000103903	0	0	0	0
Rhombic	25.4 mm	31.75 mm	0.00547024	-0.000459287	0.000013822	0	0
Square	19.05 mm	25.4 mm	-0.00409219	0.00245958	0.0003604	0.0000227548	-5.37988×10^{-7}

Table 8: Constants for the equation for the correction factor R_l

Values of S_1	A	B	C	D	E	F	G	H
0 to 0.3	0.9930511	0.04163241	-0.07862612	0.01413662	-4.092978	-6.546514	4.516072	-2.525205
0.3 to 0.8	0.7499537	-0.4381533	0.2981431	0.1471556	-0.709333	0.0847061	1.053554	-4.215857
Values of S_1	I	J	K	L	M	N	O	P
0 to 0.3	15.6087	30.3289	-23.86598	14.08358	-23.22663	-41.67332	24.92778	-15.50438
0.3 to 0.8	0.3060496	1.27137	-12.85256	19.67211	-0.212271	-3.186438	18.491314	-23.16708

Pressure Loss Correction Factor by Recirculation Effect R_b

This correction factor is provoked by circulation currents in the heat exchanger and it is computed by Equation (14) which is in terms of F_{sbp} and $N_2 = N_{ss}/N_c$.

$$R_b = \left[A + B(N_2) + C(N_2)^2 + D(N_2)^3 + E(N_2)^4 \right] + \left[F + G(N_2) + H(N_2)^2 + I(N_2)^3 + J(N_2)^4 \right] F_{sbp}$$
$$+ \left[K + L(N_2) + M(N_2)^2 + N(N_2)^3 + \tilde{N}(N_2)^4 \right] F_{sbp}^2 + \left[O + P(N_2) + Q(N_2)^2 + R(N_2)^3 + S(N_2)^4 \right] F_{sbp}^3$$
$$+ \left[T + U(N_2) + V(N_2)^2 + W(N_2)^3 + X(N_2)^4 \right] F_{sbp}^4$$

$$(14)$$

The constants demanded by the equation are defined in Table 9 and they are in terms of Re_c number. Comparing the original graph and that one obtained by Equation (14), the maximal error found is 1.3%.

Pressure Loss Correction Factor by Uneven Baffle Spacing at the Inlet and Outlet

This correction factor is proposed by the uneven baffle spacing at inlet and outlet of the heat exchanger and this factor is calculated by Equation (15).

$$R_s = \frac{1}{2} \left[\left(l_{s,i}^* \right)^{n'} + \left(l_{s,o}^* \right)^{n'} \right]$$

$$(15)$$

where $n' = 1.6$ for turbulent flow ($Re_c > 100$).

$n' = 1$ for laminar flow ($Re_c < 100$).

CALCULUS PROGRAM

Having the correction factors in an analytical form, it was developed a calculus program using the FORTRAN 77 language which will design quickly the shell and tubes heat exchanger through the improvement DELAWARE Method. This computing program is described next:

Flow Chart Program

The flow chart used for the shell and tubes heat exchanger is presented in Figure 2.

As from this flow chart, it was made the heat exchanger design program using the executable BELL since it runs in MS DOS platform.

CONCLUSIONS

This work has presented the improvements to the Delaware method for the shell and tube heat exchanger design. These improvements were made up not only of obtaining the correction factor equations which were only available in graphic forma but also of developing the computing program in Fortran 77 language.

Thus, a new and easy tool for the shell and tubes heat exchanger design is available which allows accomplishing the numerical computing in a quick form minimizing the errors of the graphical lecture. This system gives the opportunity to explore different design alternatives in order to find the optimal solution to each proposed problem.

Recent years, heat exchanger is often used for the request of technology. But the relevant design is not provided by the actual standards. This work presents the improvements to the Delaware method for the shell and tube heat exchanger design.

Table 9: Constants for the equation for the correction factor R_b

For	A	B	C	D	E	F	G	H	I
$Re_c > 100$	0.9999956	0.02427714	-0.2498571	0.8768954	-0.9389413	-3.7854133	42.7281376	-285.42201	841.836875
$Re_c < 100$	0.9989929	0.10536115	-0.8164392	2.13589149	-1.8323906	-4.7393642	60.6001915	-402.43318	1138.26259
For	J	K	L	M	N	Ñ	O	P	Q
$Re_c > 100$	-823.25082	6.9400004	-114.27156	831.289197	-2493.3366	2464.67857	-7.1165478	129.6448	-995.82525
$Re_c < 100$	-1075.7794	11.8151337	-238.94173	1739.20139	-4979.2121	4724.16651	-17.614757	423.093405	-3227.2538
For	R	S	T	U	V	W	X		
$Re_c > 100$	3071.9304	-3083.9014	3.21390213	-60.119316	480.616728	-1521.5693	1550.23124		
$Re_c < 100$	9354.34611	-8902.6809	11.6766845	-302.46692	2376.53333	-6980.2925	6687.42629		

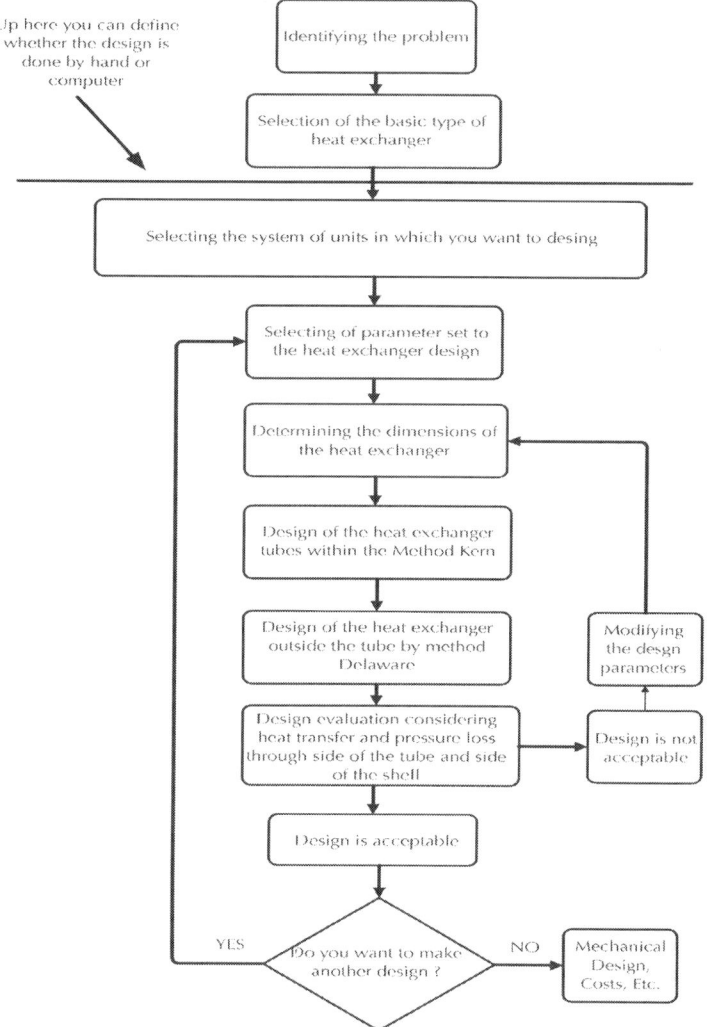

Figure 2: Flow chart program.

REFERENCES

1.   Vengateson, U. (2010) Design of Multiple Shell and Tube Heat Exchangers in Series: E Shell and F Shell. Chemical

Engineering Research and Design, 88, 725-736.http://dx.doi.org/10.1016/j.cherd.2009.10.005

2.   Costa, A.L.H. and Queiroz, E.M. (2008) Design Optimization of Shell and tube Heat Exchangers. Applied Thermal Engineering, 28, 1798-1805.http://dx.doi.org/10.1016/j.applthermaleng.2007.11.009

3.   Serna, M. and Jimenez, A. (2005) A Compact Formulation of Bell-Delaware Method for Heat Exchanger Desing and Optimation. Chemical Engineering Research and Desing, 83, 539-550. http://dx.doi.org/10.1205/cherd.03192

4.   Shah, R.K., Dusan, P. and Sekulic, D.P. (2003) Fundamentals of Heat Exchanger Design. John Wiley & Sons, Hoboken. http://dx.doi.org/10.1002/9780470172605

5.   Ayub, Z.H. (2005) A New Chart Method for Evaluating Single-Phase Shell Side Heat Transfer Coefficient in a Single Segmental Shell and Tube Heat Exchanger. Applied Thermal Engineering, 25, 2412-2420.http://dx.doi.org/10.1016/j.applthermaleng.2004.12.015

6.   Leong, K.C., Toh, K.C. and Leong, Y.C. (1998) Shell and Tube Heat Exchanger Design Software for Educational Applications. International Journal of Engineering Education, 14, 217-224

7.   León, A.R., Velazquez, M.T. and Diez, P.Q. (2011) The Design of Heat Exchangers. Engineering, 3, 911-920.

8.   Castillo, R. (1999) Diseño Computacional de Intercambiadores de Calor de Coraza y Tubos por el Método Delaware. Tesis de Maestría, Instituto Politécnico Nacional, México.

Chapter 3

Performance Assessment of A Shell Tube Evaporator For A Model Organic Rankin Cycle For Use In Geothermal Power Plant

Haile Araya Nigusse, Hiram M. Ndiritu, and
Robert Kiplimo

Department of Mechanical Engineering, Jomo Kenyatta University
of Agriculture and Technology, Nairobi, Kenya

ABSTRACT

The global energy demand increases with development and population rise. Most electrical power is currently generated by conventional methods from fossil fuels. Despite the high energy demand, the conventional energy resources such as fossil fuels have been declining and harmful combustion byproducts are causing global warming. The Organic Rankine Cycle power plant is a very effective option for utilization of low grade heat sources for power generation. In the Organic Rankine Cycle heat exchangers such as evaporators and condensers are key components that determine its performance. Researches indicated that shell tube heat exchangers are effectively utilized in this cycle. The design of the heat exchanger involves establishing the right flow pattern of the interacting fluids. The performance of these exchangers can be optimized by inserting baffles in the shell to direct the flow of fluid across the tubes on shell side. In this work heat exchangers have been developed to improve heat recovery from geothermal brine for additional power generation. The design involved sizing of heat exchanger (evaporator) using the LMTD method based on an expected heat transfer rate. The heat exchanger of the model power plant was tested in which hot water simulated brine. The results indicated that the heat exchanger is thermally suitable for the evaporator of the model power plant.

INTRODUCTION

Energy consumption increases with growth in population. The increase in energy demand and high cost of fossil fuels is the main challenge for the developing countries. Furthermore, the increase of environmental concerns and energy crisis has resulted in need for a sustainable approach to the utilization of the earth's energy resources. Geothermal energy is the energy derived from the natural heat of the earth. The heat from the earth's own molten core is conducted to the adjacent rocks and transferred to underground water reservoirs by convection. The steam (hot water) heated by the geothermal heat can be tapped using different technologies. Geothermal energy is among the most reliable forms of renewable energy. Most of the other clean energy sources are weather dependent [1] .

Worldwide, geothermal power plants have a capacity of about

12 GW power generations as of 2013 and in practice supply only about 0.3% of global power demand [2] . Moreover, the conventional geothermal power plants utilize the high temperature and pressure geo-fluid and operate at low efficiency due to the heat loss in the exhausted steam and brine.

Despite the simplicity and least cost, flash steam geothermal power plants operate at lower efficiency mainly due to the untapped available energy in the brine. Discharged geothermal brine generally has a temperature higher than 100°C and mass flow rate of hundreds of tons/hour [3] . This rejected high temperature brine wastes significant amount of available energy and causes thermal pollution to the environment. Organic Rankine Cycle power plant is an advantageous technology that consents the power generation from low temperature water dominant geothermal resource and energy recovery from the geothermal brine.

The lab scale model Organic Rankine Cycle (ORC) power plant was designed to operate at a heat source temperature of 50°C to 60°C and pressure between 1 and 2.5 bars to simulate the energy recovery from the low temperature geothermal brine discharged at the reservoir of flash steam geothermal power plant. The components of a model Organic Rankine Cycle power plant for recovery of waste heat from the geothermal brine to increase power generation were fabricated and performance assessment of each component was conducted. The model power plant utilizes hot water simulating the geothermal brine and comprises of the evaporator, turbine, condenser and feed pump. Heat exchangers are the important component linking the geo-fluid and ORC.

In the cycle shown in Figure 1, the feed pump supplies pentane from the working fluid tank to the evaporator, where the working fluid is vaporized by heat transferred from the hot water. The high pressure vapor from the evaporator flows into the turbine and expands producing mechanical power. The low pressure vapor exhausted from the turbine is condensed by cooling water as it flows through the condenser. The condensed working fluid is directed back into the reservoir and pumped again into the evaporator and a new cycle began. Figure 1 is the T-S diagram of the cycle and shows that the working fluid pentane is a dry expansion. The saturation curve is skewed to the right and shows expansion ends on the dry region on which no condensation happens in the turbine.

The most commonly used heat exchangers in ORC are the shell tube evaporator and condenser. Their widespread application can be attributed to ease of manufacture from a variety of materials. Moreover, there is no limit on the operating temperature and pressure [4] [5] . The heat exchangers are built in accordance with three mechanical standards that specify design, fabrication, and materials of unfired shell and tube heat exchangers. The two most common heat exchanger design problems are those of sizing and rating of the heat exchanger [6] .

Shell and tube heat exchangers are non-compact exchangers. Generally the heat transfer surface area per unit volume is high ranging from about 50 to 100 m²/m³. Thus, they require a considerable amount of space, support ing structure, and higher capital and installation costs. The shell and tube heat exchangers are more effective for applications in which compactness is not a priority [6] .

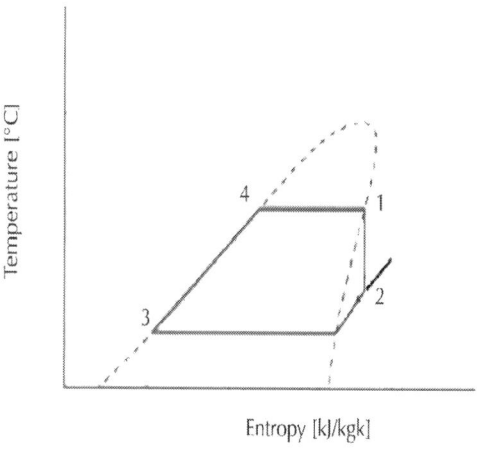

Figure 1: The T-S diagram of the cycle.

Madhawa [7] investigated the flat plate and shell tube heat exchangers and found that flat plate type heat exchangers were effective in the evaporator and condenser when considering low-temperature heat sources in which large heat exchanger area (per unit power output) was required. The plate type heat exchangers are preferred due to their compactness and high heat transfer co-efficient which result to less heat transfer area than would be needed using shell and tube heat exchanger. However, flat plate exchanger involves high manufacturing cost and maintenance cost which would affect the overall cost of the power plant.

Bambang [8] also compared the flat plate and shell tube heat exchangers and found that the shell tube type was advantageous due to simplicity in geometry, well established design procedure, could be constructed from a wide range of materials, used well-established fabrication techniques and was easily cleaned. However, the heat transfer optimization is a key challenge mainly due to large pressure drops within the shell tube heat exchangers.

The design of heat exchangers involves establishing the right flow pattern of the interacting fluids. Parallel and counter flows are the two common flow patterns in shell tube heat exchanger. The counter flow is the predominantly preferred flow direction in liquid to liquid heat exchangers since it results in a higher temperature difference [9] .

Various studies have been carried out on optimization of the performance of shell and tube heat exchangers using the performance parameters approach. The heat transfer coefficient values are evaluated using the log mean temperature difference (LMTD) method from the temperature difference and the heat transfer area for known inlet and outlet temperature heat exchangers [10] . Thundil et al. [11] investigated the effect of inclination of baffles in the shell by simulating a model shell and tube heat exchanger. This involved comparing the impact of baffle inclination on fluid flow, pressure drop and the heat transfer characteristics of a shell tube heat exchanger using three different inclination angles (0°, 10° and 20°). They concluded that shell and tube heat exchanger with 20°

baffle inclination angle resulted in better performance compared to 10° and 0° inclination angles.

A. Singh et al. [12] analyzed on the performance of a shell tube heat exchanger with segmented baffles at three different orientations (0°, 30° and 60°). They analyzed the system for laminar flow with varying Reynolds number and concluded that the heat transfer coefficient increased with increase in Reynolds number in shell tube heat exchanger for both hot fluid and cold fluids. They observed that, with the introduction of the baffles, the heat transfer coefficient increased leading to more heat transfer rate due to introduction of swirl and more convective surface area. This means that baffles are necessary components that improve performance of heat exchangers.

In general, conventional shell tube heat exchangers result in high shell-side pressure drop and formation of re-circulation zones near the baffles. To overcome the challenge, helical baffles, which give better performance than single segmental baffles, can be used. But these baffles involve high manufacturing cost, installation cost and maintenance cost. Hence the effectiveness and cost must be considered in the heat exchanger design [11] .

This paper presents design, fabricate and performance assessment of the shell and tube heat exchanger designed for the evaporator of a model Organic Rankine Cycle power plant. Performance tests were conducted by relating the inlet and outlet temperatures and the overall heat transfer coefficient to the rate of heat transfer between the two fluids.

METHODOLOGY

Heat Exchanger Selection

Heat is transferred from one fluid to other in the heat exchanger. Heat transfer area, overall heat transfer coefficient and temperature difference are important factors to be considered [13] . For indirect heat transfer between two fluids, the shell and tube heat exchangers are more effective. In this study, a shell and tube type heat exchanger is selected for the evaporator of the model power plant due to the advantages of fairly simple geometry, can be fabricated from a wide range of materials and ease of cleaning.

The design of heat exchangers involves establishing the right flow configuration of heat exchangers. Parallel flow and counter flow are the main flow configurations in shell tube heat exchanger. The counter flow configuration is the predominantly preferred flow direction in liquid to liquid heat exchangers since it results in a higher temperature difference driving the heat transfer within the heat exchanger, smaller heat transfer surface area is required [9] . Moreover counter flow configuration is most effective design when the desired outlet temperature of secondary fluid is between the inlet and outlet temperatures of the hot water [14] .

The design of the evaporator is a fixed tube counter flow shell tube heat exchanger as shown inFigure 2. Moreover, the fluid flow in the exchanger is in such way that the hot water flows in the tube side and the secondary fluid (n-pentane) vaporizes in the shell side of the evaporator. This type of fluid flow configuration allow fouling fluid to flow through the tubes (easier to clean) and the organic working fluid (n-pentane) flows through the shell side where a turbulent flow is obtained due the baffles. Turbulence eddies are induced due to recirculation near the baffles and which would result in more pressure drop with 0° baffle orientation [11] .

Pentane has been selected for the working fluid of the model power plant due to its convenience for thermodynamic cycle working between normal atmospheric condition and the boiling condition of water. Moreover pentane is environmentally friendly (non-ozone depleting) and lower global worming potential organic fluid. Pentane is liquid state at atmospheric conditions (minimizes handling cost) and has good performance for ORC. The thermodynamic properties of pentane are acquired from a database program called NIST WebBook (NIST standard Reference Database Number 69) [15] .

Design and Construction of the Evaporator

In design of heat exchangers it is reasonable to assume a constant value of overall heat transfer coefficient (U). The log mean temperature difference (LMTD) method is useful for the sizing and rating of heat exchanger of known mass flow rate and range of temperature difference between the inlet and outlet of fluid streams shown in Table 1. The following equations have been used to relate the heat transfer surface area through which heat flow occurs under the driving force of temperature difference, amount of heat transferred and overall heat transfer coefficient.

$$Q_{hot} = C_h \left(T_{hi} - T_{ho} \right) \tag{1}$$

where heat capacity rate for hot or cold fluid, $C = mC_p$. Where C_p is specific heat capacity at constant pressure and m is the mass flow rate of working fluid.

During the heat transfer process energy is conserved therefore for the cold and hot fluid;

$$Q_{hot} = Q_{cold} = C_h \left(T_{hi} - T_{ho} \right) = C_c \left(T_{co} - T_{ci} \right) \tag{2}$$

where heat capacity rate for hot or cold fluid $C = mc_p$, subscripts h_i and h_o represent inlet and outlet of hot fluid respectively and c_i and c_o represent cold fluid inlet and outlet respectively.

Heat transferred in the process (Q) may be related to the overall heat transfer coefficient U and the mean temperature difference ΔT_{LMTD} by:

$$Q = AUT_{LMTD} \tag{3}$$

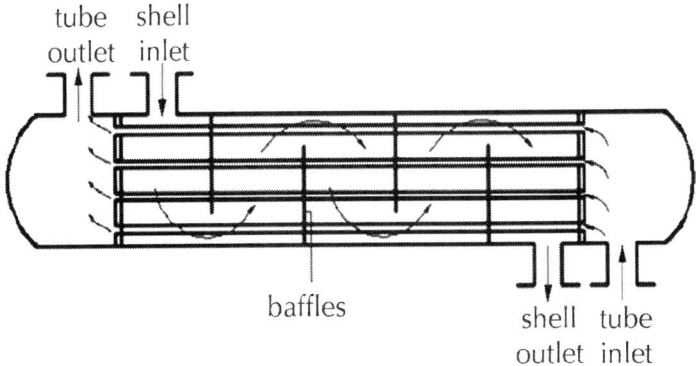

Figure 2: Counter flow shell and tube heat exchanger [4] .

Table 1: Flow conditions of the fluid streams in the heat exchanger

Sr. No	Description	Tube side	Shell side
1	Mass flow rate (kg/s)	0.1	0.12
2	Inlet temperature (°C)	54	25
3	Outlet temperature (°C)	36	51

where A is heat transfer surface area and ΔT_{LMTD} is the log mean temperature difference. The log mean temperature difference of the heat exchanger can be determined as follows:

$$T_{LMTD} = \frac{(\Delta T_1 - \Delta T_2)}{\ln\left[\frac{(\Delta T_1)}{(\Delta T_2)}\right]} = \frac{(T_1 - T_4) - (T_2 - T_3)}{\ln\left[\frac{(T_1 - T_4)}{(T_2 - T_3)}\right]}$$

(4)

where, ΔT_1 and ΔT_2 represent the temperature differences at the inlet and exit of the heat exchanger respectively as indicated in Figure 3.

The heat exchanger was designed using the design Equations (1)-(4). The details of the designed shell and tube heat exchanger are shown in Table 2. Figure 4 shows the assembly of the copper tubes to the tube sheet and the baffles placed on the respective position.

Evaluation of the thermal parameters:

The following relations were applied to evaluate the remaining performance parameters of the heat exchanger. The LMTD method was used since the inlet and outlet temperatures of both the hot water and the secondary fluid are known.

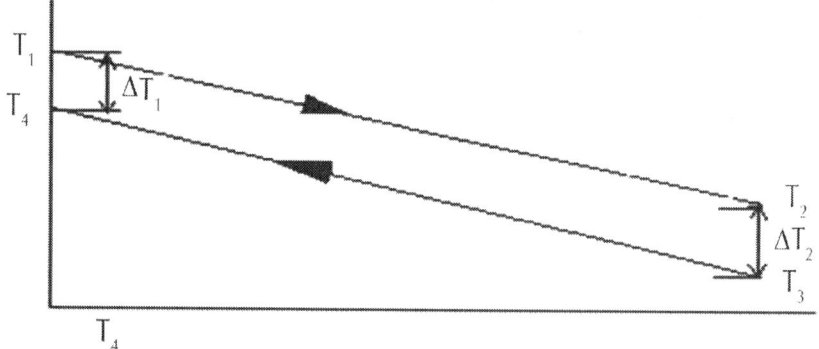

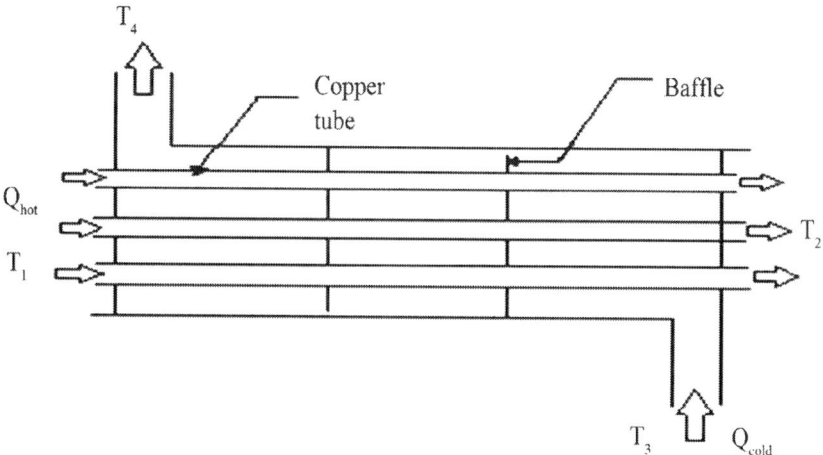

Figure 3: Flow configuration of the evaporator.

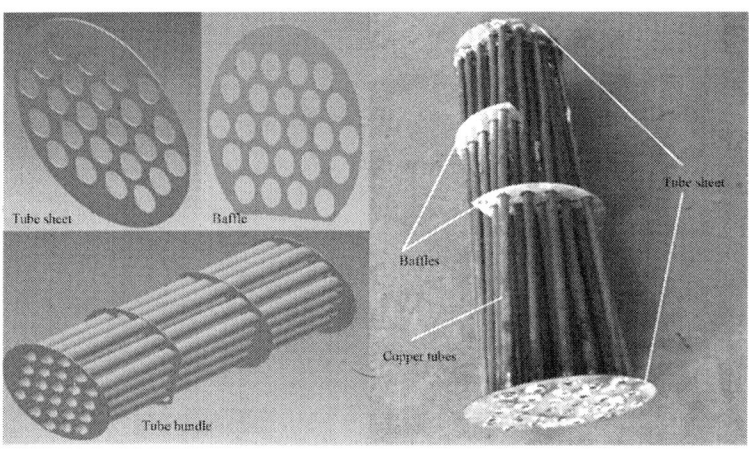

Figure 4: The shell and tube heat exchanger.

Table 2: Dimensions and parameters of the heat exchanger

Sr. No	Description	Tube side	Shell side
1	Interacting fluids	Hot water	Pentane
2	Number of tubes	33	
3	Total tube length (m)	13.1	
4	Tube diameter (outer/ inner) (mm)	12.7/10.9	150
5	Heat transfer area (m²)	0.41	
6	Number of pass	1	
7	Tube configuration	30° triangular	
8	Tube pitch (mm)	16	
9	Material	Copper	Galvanized steel

1) Heat duty (Q)

$$Q_{hot} = C_h \left(T_{hi} - T_{ho} \right) = A U T_{LMTD}$$

(5)

$$Q_{cold} = C_c \left(T_{co} - T_{ci} \right) = m C_p \Delta T \tag{6}$$

where subscripts c and h represent cold and hot fluid respectively, i and o represent for inlet and outlet respectively. Heat capacity rate for hot or cold fluid $C = m C_p$, A is the heat transfer area and U is overall heat transfer coefficient and T_{LMTD} is the log mean temperature difference.

- Cooling water pressure drop, $\Delta P_w = P_i - p_o$
- Pressure drop of pentane, $\Delta P_p = P_i - P_o$
- Cooling water temperature difference, $\Delta T_w = T_{wi} - T_{wo}$
- Pentane temperature difference, $\Delta T_p = T_{pi} - T_{po}$
- Log mean temperature difference T_{LMTD} for counter flow heat exchanger:

$$T_{LMTD} = \frac{\left(T_{wi} - T_{po} \right) - \left(T_{wo} - T_{pi} \right)}{\ln \left[\dfrac{\left(T_{wi} - T_{po} \right)}{\left(T_{wo} - T_{pi} \right)} \right]}$$

where:

T_{wi} is cooling water inlet temperature; T_{wo} is cooling water outlet temperature, T_{pi} is pentane inlet temperature, and T_{po} is pentane outlet temperature.

Fabrication of the Heat Exchanger

The shell of the heat exchanger was constructed by rolling a plate of galvanized steel. The plate was cut to a rectangular shape of size 471 mm by 750 mm and four holes were drilled for the inlet and outlet of the two fluids before rolling the plate. After rolling a shell of 150 mm diameter by 750 mm length was obtained. The copper tubes were held at the ends by means of galvanized iron sheets as shown in Figure 4. Two baffles were also provided as shown with a baffle pitch of 150 mm. The optimum baffle pitch (spacing between segmental baffles) and the baffle cut were used to determine the cross flow velocity and hence the rate of heat transfer and minimize pressure drop. A baffle spacing of 0.2 to 1 times of the inside shell diameter is commonly used [16] .

Copper tubes were cut at a length of 400 mm and each tube was brazed using a gas welding to the hole provided on the tube sheet to produce a tube bundle as shown in Figure 4. The tube bundle then inserted to the main shell and the two tube sheets were brazed using gas welding to provide air tight joints.

Heat Exchanger Leakage Test

It is important to ensure there is no leakage of fluids of the heat exchanger at the operating temperature and pressure before carrying out a performance test. The heat exchanger was tested for leakages using water both for the shell side and tube side. Water pumped at a pressure of 2.5 bars was used for this test. Any leaking sections (particularly at the welded joints) were sealed using a sealing material called STAG. STAG is easy break, smooth consistency and lead-free joining compound. It was applied to the evaporator since it is non-poisonous and does not react with pentane and water.

Performance Test of the Heat Exchanger

The performance test unit consists of an overhead hot water tank, the heat exchanger, secondary fluid tank and feed pump. A schematic sketch showing valves, pressure gauges, flow meters, and the location of temperature sensors is given in Figure 5. The hot water, produced by an electric heater in the hot water tank, flows by gravity through the heat exchanger in the tube side. Hot water flow rate is controlled by opening a flow valve HV1.

The feed pump supplies the secondary fluid (n-pentane) to the exchanger through the flow meter (FM1) and evaporates on the shell side in a counter-current to the hot water flow. The vapor which evaporated in the heat exchanger is then directed back to the secondary fluid tank through the condenser.

The length of one test section is set to be two minutes. A set of four thermocouples are provided to record pertinent temperatures at the inlet and exit of each fluid as shown in the Figure 5. The thermocouples are connected to a data logger called TDS-530. The data logger was set to print each temperature in a test section of two minutes. A set of five measurements was taken by varying the hot water mass flow rate and maintaining the secondary fluid (n-pentane) flow rate constant. The temperature indicators on the data logger displayed the temperatures of hot water at the inlet (T_1), hot water at the exit (T_2), secondary fluid at the inlet (T_3) and secondary fluid at the exit (T_4) of the heat exchanger.

Experimental Procedures

The overhead hot water tank was filled with water. The three kilowatt water heater built in the hot water tank was switched on and temperature was set to 54°C. The inlet valve HV1 was opened and hot water was allowed into the tube side of the heat exchanger. The feed pump was switched on and the secondary fluid (n-pentane) allowed to flow into the shell side of the heat exchanger. A steady state of the measurements was attained in 36 minutes and all the four temperatures and flow rates of hot water and secondary fluid was recorded. The flow rate of hot water was changed and waited for new steady state to be reached.

RESULTS AND DISCUSSION

Performance test of the heat exchanger was carried out. Three parameters were studied to understand the performance of the heat exchanger. The heat transferred, overall heat transfer coefficient, tube side and shell side pressure drop within the exchanger. The shell tube heat exchanger was having hot water in tube side and secondary fluid (n-pentane) in the shell side in counter flow configuration. Five experiments were conducted at different hot water mass flow rates.

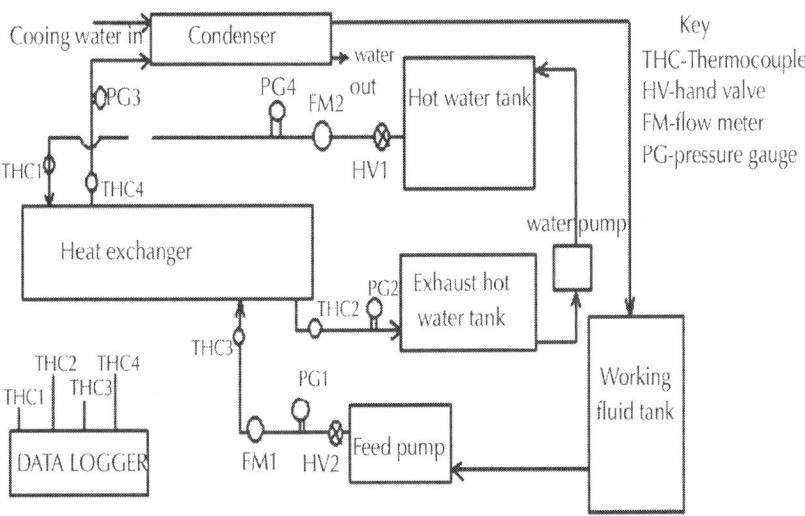

Figure 5: Schematic sketch of the evaporator performance assessment experiment.

Increase in the flow rate of hot water resulted in increase in the overall heat transfer coefficient as can be seen from curve in Figure 6. The curve shows that an increase of hot water flow rate from 0.16 kg/s to 0.24 kg/s increased the overall heat transfer coefficient of the heat exchanger by 17.33%. This is because increase in the mass flow rate of hot water increases the heat energy transferred. Since the specific heat remains almost constant, hot water outlet temperature should decrease to comply with law of conservation of energy and hence as the flow rate of the hot water is increased, the tube side overall heat transfer coefficient also increases. The performance test result indicated that the developed shell and tube heat exchanger performs satisfactorily under standard conditions and the variation of the overall heat transfer coefficient and total heat transferred with the mass flow rate of the hot water is analogous to similar heat exchangers.

Similarly, the variation of heat transferred with mass flow rate of the hot water is shown in Figure 7. The curve shows that heat transferred increased by 6.74% with an increase of hot water flow rate from 0.16 kg/s to 0.24 kg/s. This is because increase in the hot water mass flow rate increases overall heat transfer coefficient in a faster rate than the heat energy transferred.

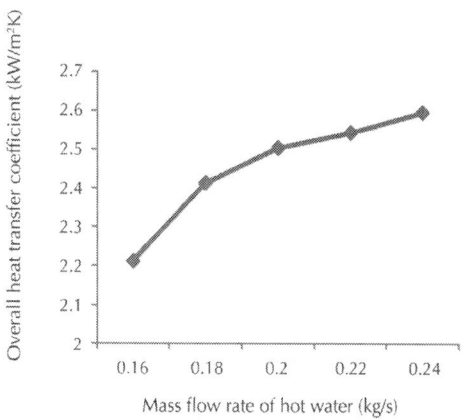

Figure 6: Variation of the overall heat transfer coefficient with increase in hot water mass flow rate.

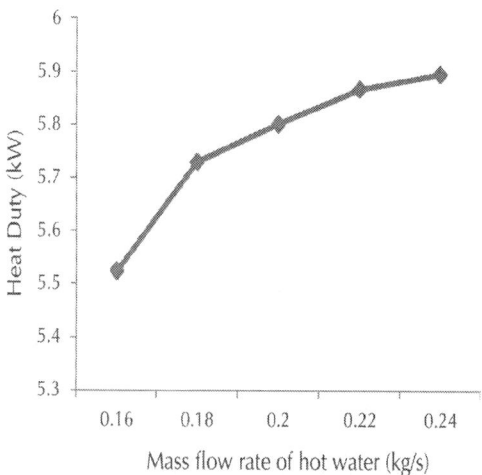

Figure 7: Variation of heat duty with increase in hot water mass flow rate.

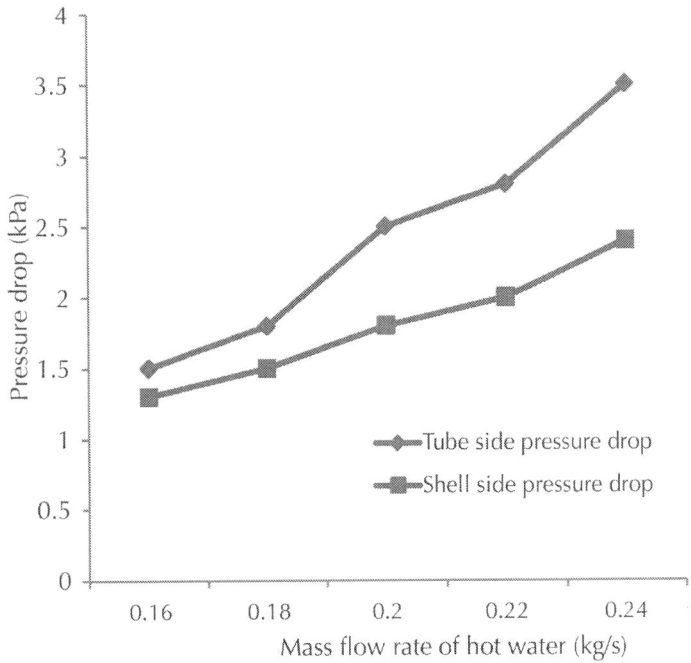

Figure 8: Variation of the pressure drop with increase in hot water mass flow rate.

Figure 8 shows the variation of tube side and shell side pressure drop values with increase in hot water mass flow rate. It was observed that the pressure drop increased by more than twice in the tube side and in the shell side pressure drop increased by 84.6% with increase in hot water flow rate in heat exchanger. Shell and tube heat exchangers generally experience pressure drop mainly due to friction, change in thermodynamic properties like viscosity and density through the heat exchanger as a result of heating or cooling, acceleration and deceleration of fluid with change is flow cross section.

This pressure drop may increase pumping power and may affect the service time of structural components of the heat exchanger. However compared to rate of change of the heat transferred the rate of increase in pressure drop is reasonable.

CONCLUSIONS

Shell and tube heat exchanger was designed and fabricated. Experimental test was conducted to study the performance of the heat exchanger. The performance parameters, the overall heat transfer coefficient, heat transferred and tube side pressure drop and shell side pressure drop at different hot water mass flow rates were evaluated. The main points are summarized as follows.

- The results of the performance test indicated that the overall heat transfer coefficient was greater than the assumed overall heat transfer coefficient of heat exchanger. This implies that the heat exchanger is thermally suitable for the evaporator of the model power plant.

- The results of the performance test revealed that the heat exchanger was working satisfactorily under standard conditions. The test results were compared with the design data and the performance parameters reasonably close to the design performance data. Thus the heat exchanger is reliable and can be applied for the evaporator of the model Organic Rankine Cycle power plant.

REFERENCES

1.   Mburu, M. (2009) Geothermal Energy Utilization. Exploration for Geothermal Resources.

2.   (2013) 2013 Geothermal Power: International Market Overview.

3.   Teguh, P.B., et al. (2011) Design of n-Butane Radial Inflow Turbine for 100 kW Binary Cycle Power Plant. International Journal of Engineering & Technology, 11, 55.

4.   Hadidi, A., Hadidi, M. and Nazari, A. (2013) A New Design Approach for Shell-and-Tube Heat Exchangers Using Imperialist Competitive Algorithm (ICA) from Economic Point of View. Energy Conversion and Management, 67, 66-74. http://dx.doi.org/10.1016/j.enconman.2012.11.017

5. Sairam, V., Siddarath, B., Saiprasad, C., Taruns, S. and Sujit, G. (2014) Design, Fabrication and Testing of FRP Shell Counter-Flow Heat Exchanger. International Journal of Engineering Science and Innovative Technology, 3, 171-176.

6. Shah, R.K. and Sekulic, D.P. (2012) Fundamentals of Heat Exchanger Design. University of Kentucky, Lexington.

7. Hettiarachchi, H.D.M., et al. (2007) Optimum Design Criteria for an Organic Rankine Cycle Using Low-Temperature Geothermal Heat Sources. Energy, 32, 1698-1706.http://dx.doi.org/10.1016/j.energy.2007.01.005

8. Teguh, P.B. and Trisno, M.D. (2011) Model of Binary Cycle Power Plant Using Brine as Thermal Energy Sources and Development Potential in Sibayak. International Journal of Electrical & Computer Sciences, 11, 45.

9. Kuppan, T. (2000) Heat Exchanger Design Handbook. Marcel Dekker, Inc., New York.

10. Incropera, F.P. and Dewitt, D.P. (2006) Introduction to Heat Transfer. New York.

11. Thundil, R. (2012) Shell Side Numerical Analysis of a Shell and Tube Heat Exchanger Considering the Effects of Baffle Inclination Angle on Fluid Flow. International Journal of Thermal Science, 16, 1165-1174. http://dx.doi.org/10.2298/TSCI110330118R

12. Singh, A. and Sehgal, S.S. (2013) Thermohydraulic Analysis of Shell-and-Tube Heat Exchanger with Segmental Baffles. ISRN Chemical Engineering, 2013, Article ID: 548676.

13. Kumar, U., Karimi, M.N. and Agrawal, B.K. (2013) Optimization and Selection of Organic Rakine Cycle for Low Grade Heat Recovery by Using Graph Theoretical Approach. International Journal of Sustainable Development and Green Economics, 2, 11-16.

14. Pandey, A. (2011) Performance Analysis of a Compact Heat Exchanger. Master's Thesis, National Institute of Technology Rourkela, Rourkela.

15. NIST Chemistry WebBook. http://webbook.nist.gov/chemistry/

16. Patel, S.K. and Mavani, A.M. (2012) Shell and Tube Heat Exchanger Thermal Design with Optimization of Mass Flow Rate and Baffle Spacing. International Journal of Advanced Engineering Research and Studies, 2, 130-135.

Chapter 4

Design and Development of Shell & Tube Heat Exchanger for Beverage

Shravan H. Gawande[1], Sunil D. Wankhede[1], Rahul N. Yerrawar[1], Vaishali J. Sonawane[1], and Umesh B. Ubarhande[2]

[1]Department of Mechanical Engineering, M. E. Society's College of Engineering, Pune, India

[2]Dy. Manager, PEM Design (R&D), Alfa Laval (India) Ltd., Pune, India

ABSTRACT

In the first part of this paper, a simplified approach to design a Shell & Tube Heat Exchanger [STHE] for beverage and process industry application is presented. The design of STHE includes thermal design and mechanical design. The thermal design of STHE involves evaluation of required effective surface area (i.e. number of tubes) and finding out log mean temperature difference [LMTD]. Whereas, the mechanical design includes the design of main shell under internal & external pressure, tube design, baffles design gasket, etc. The design was carried out by referring ASME/TEMA standards, available at the company. The complete design, fabrication, testing and analysis work was carried out at Alfa Laval (India), Ltd., Pune-12. In the second part of this paper detail view of design optimization is presented by flow induced vibration analysis [FVA].

INTRODUCTION

Heat Exchangers are devices used to enhance or facilitate the flow of heat. Every living thing is equipped in some way or another with heat exchangers. They are widely used in space heating, refrigeration, air conditioning, power plants, chemical plants, petrochemical plants, petroleum refineries, natural gas processing, and sewage treatment. The design of STHE including thermodynamic and fluid dynamic design, cost estimation and optimization, represents a complex process containing an integrated whole of design rules and empirical knowledge of various fields.

The design of STHE involves a large number of geometric and operating variables as a part of the search for heat exchanger geometry that meets the heat duty requirement and a given set of design constrains. A STHE is the most common type of heat exchanger in oil refineries and other large chemical processes, and is suited for higher-pressure applications. As its name implies, this type of heat exchanger consists of a shell (a large vessel) with a bundle of tubes inside it. One fluid runs through the tubes and the second runs over the tubes (through the shell) to transfer heat between the two fluids. A set of tubes is called a tube bundle which may be composed by several types of tubes e.g. plain, longitudinally finned, etc.

EXISTING INDUSTRIAL SCENARIO

In industries, heat exchangers are used in industrial process to recover heat between two process fluids. Shell-and-tube heat exchangers are the most widely used heat exchangers in process industries because of their relatively simple manufacturing and their adaptability to different operating conditions. But nowadays numbers of industries are searching for effective and less time consuming alternatives of designing of shell-and-tube heat exchangers. As per literature and industrial survey it is observed that there is need of effective design options for STHE. This section explains the details of existing industrial scenario of design of STHE.

Part A-Thermal Design

The thermal design of STHE includes:

- Consideration of process fluids in both shell and tube side;
- Selection of required temperature specifications;
- Limiting the shell and tube side pressure drop;
- Setting shell and tube side velocity limits;
- Finding heat transfer area including fouling factor.

Part B-Mechanical Design

The mechanical design of STHE includes:

- Selection of TEMA layout—based on thermal design;
- Selection of tube parameters such as size, thickness, layout, pitch, material;
- Limiting the upper and lower design on tube length;
- Selection of shell side parameters such as material, baffle spacing, and clearances;
- Thermal conductivity of tube material;
- Setting upper and lower design limits on shell diameter and baffle spacing.

As per literature and industrial survey at [A₁ & A₂] the design is carried out using in-house developed software for design and drafting. This dedicated software enables qualified engineers to accomplish complex design calculations complying strictly with the requisite international codes and standards. The software also generates fabrication drawings to scale and 3-D images of the Exchanger thereby giving warning of any foul-up/mis-match in nozzles, RF-Pads and in the dimensions of various components. Also an experienced team of design engineers undertakes thermal and mechanical design of complex heat exchangers and generate fabrication drawings to scale along with weights and estimates based on customer's specifications. These designs are optimized to arrive at an optimal size. After carrying out the design, the final output is in an AutoCAD drawing format (DWG) or DWF (Web format).

In this proposed work design, development & testing of STHE is carried out. Along with the parameter considered as per [A1 & A2], the software generated design was cross checked with manual design. Also vibration analysis is performed to optimize unsupported span of tube by using HTRI software.

The paper is organized as; detailed overview on work carried out by researchers is presented in section 1 and section 2 presents the existing industrial scenario of STHE design, section 3 states the current problem definition & objective, detailed design (thermal and mechanical design) and details of STHE are given in section 4. Section 5 explores manufacturing of STHE, while section 6 describes hydraulic testing of STHE and concluding remark is given in section 7.

PROBLEM DEFINITION & OBJECTIVE

The problem presented in this paper is to design & develop a STHE, conforming to the TEMA/ASME [1-2] Standards, based on following Input Data:

1) Inlet & Outlet Temperatures of fluids on Shell & Tube Side2) Tube length = 10,000 mm3) Tube OD = 38.1 mm4) Shell OD = 1350 mm.

As per the requirement the objective of the preset work is to perform thermal and mechanical design of STHE using TEMA/ ASME standards to reduce time.

DESIGN OF STHE

The design of STHE involves a large number of geometric and operating variables as a part of the search for an exchanger geometry that meets the heat duty requirement and a given set of design constrains. Usually a reference geometric configuration of the equipment is chosen at first and an allowable pressure drop value is fixed. Then, the values of the design variables are defined based on the design specifications and the assumption of several mechanical and thermodynamic parameters in order to have a satisfactory heat transfer coefficient leading to a suitable utilization of the heat exchange surface. The designer's choices are then verified based on iterative procedures involving many trials until a reasonable design is obtained which meets design specifications with a satisfying compromise between pressure drops and thermal exchange performances [3-10].

The details of shell and tube heat exchanger under consideration are shown in Table 1.

The details of STHE are shown in Table 1. Figure 1 shows the various major components of a typical STHE as listed below:

- Connections
- Tube Sheet
- Gasket
- Head/Dish End
- Mounting/Support
- Baffles

- Shell
- Tube Bundle

Table 1: STHE specification

Parameter	Description
Size (Dia./length)	Ø1336/10,000 mm
Surface area (eff.)/unit	781.4 m. sq.
Shells/unit	1
Heat exchanged, (Q)	5064.9 KW
LMTD (Corrected)	9.15°C

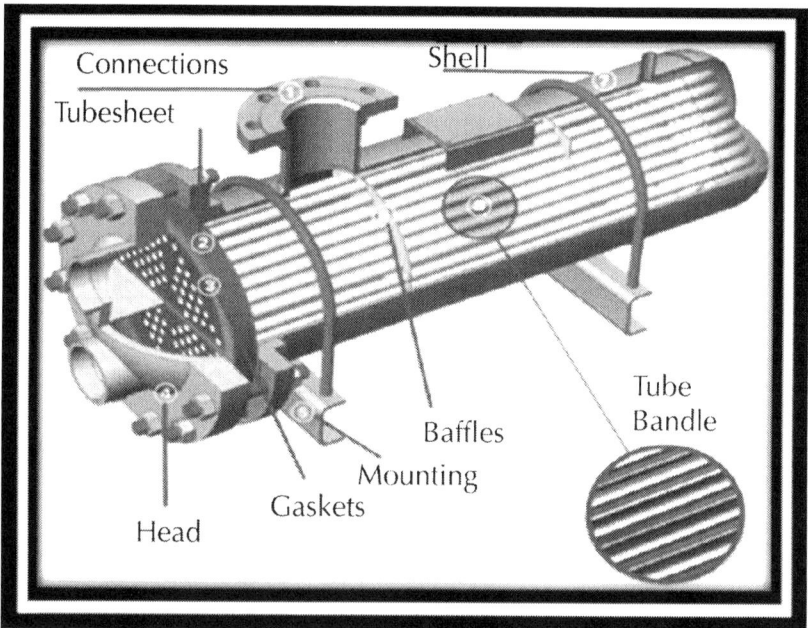

Figure 1: Major components of a typical STHE.

Part-A: Thermal Design

Design input:
- Mass flow rate (m)
- Heat exchanged (Q)
- Shell side inlet temperature (t_{si})
- Shell side outside temperature (t_{so})
- Tube side inlet temperature (t_{ti})
- Tube side outlet temperature (t_{to})
- Transfer rate (U)
- Tube outside diameter (d_o)
- Length of tube (L)

Output:
- LMTD (Log Mean Temp. Difference) (θ_m):

$$\theta_m = (\theta_1 - \theta_2)/(\ln(\theta_1/\theta_2))$$
$$= (t_{si} - t_{ti}) - (t_{so} - t_{to})/\ln((t_{si} - t_{ti})/(t_{so} - t_{to}))$$
$$= 9.15\,C.$$

- Area (A):

$$Q = U \times A \times \theta_m \quad \text{where,} \quad A = 781.61 \text{ m}^2.$$

- Number of Tubes (n):

$$Q = U \times n \times \pi \times d_o \times L \times \Delta_t \quad \text{where,} \quad n = 660 \text{ tubes.}$$

Part-B: Mechanical Design

Main Shell Design

Design Input:
- Internal design pressure (P_s)
- External design pressure (P_e)
- Shell outside diameter (D)

- Joint efficiency for longitudinal joint (K)
- Joint efficiency for circumferential joint (E)
- Allowable Stress (S)
- Element thickness (t)
- Joint efficiency for outer longitudinal joint (K_o)
- Shell outside diameter with allowance (D_{so})
- Constant Factor (B)

Output:

Design under Internal Pressure:

- Required thickness due to internal pressure (t_r):

$$= \left(P_s \times D \times K\right)/\left(2 \times S \times E - 0.2 \times P_s\right) = 4.92 \text{ mm}$$

- Max. allowable working pressure at given thickness

$$= \left(2 \times S \times E \times t\right)/\left(K \times D + 0.2 \times t\right) = 7.99 \text{ bar}$$

- Max. allowable pressure, new and cold

$$= \left(2 \times S \times E \times t\right)/\left(K \times D + 0.2 \times t\right) = 10 \text{ bar}$$

- Actual stress at given pressure and thickness

$$= \left(P_s \times R\right)/\left(S \times E - 0.6 \times P_s\right) = 4.94 \text{ N/mm}^2$$

- Straight flange maximum allowable working pressure:

$$= \left(S \times E \times t\right)/\left(R + 0.6 \times t\right) = 9.27 \text{ bar}$$

Design under external pressure:

- Maximum allowable external pressure (MAEP):

$$\text{MAEP} = B/\left(K_o \times D_{so}/t\right) = 2.76 \text{ bar}$$

Tube Design

Design input:

- Internal Design Pressure (P_t)
- Allowable stress at design Temperature (S)
- Outside diameter (D_o)

- Joint efficiency for Longitudinal joint (E)

Output:

- Required Tube Thickness:

$$T_{rt1} = (P_i \times D_o)/(2(S \times E + 0.4 \times P_i)) = 1.24 \text{ mm}$$

- Baffles and Spacing
- Baffle type: Triple segmental
- Nominal shell ID: 1336 mm
- Baffle spacing (Min.): segmental baffles should not be spaced closer than 1/5th of the shell ID or 2" (51 mm), whichever greater. However, special design considerations may dictate a closer spacing.
- Baffle spacing (center to center) = 570 mm
- Spacing at inlet = 670.67 mm
- Baffle thickness = 12.7 mm

Tube Pitch

Tubes shall be spaced with minimum center to center distance of 1.25 times the outside diameter of the tube. When mechanical cleaning of the tubes is specified by the purchaser and the nominal shell diameter, minimum cleaning lanes of 1/4" (6.4 mm) shall be provided.

Gasket

The minimum width of peripheral ring gaskets for external joints shall be 3/8" (9.5 mm) for shell sizes through 23" (584 mm) nominal diameter and 1/2" (12.7 mm) for all larger sizes.

MANUFACTURING OF STHE

In this section detail of manufacturing of STHE is explained as shown in Figures 2-5.

Fabrication of Main Shell and Channel Shell

The steps in fabrication of main shell and channel shell are stated as below;

- Material identification
- Marking
- Punching
- Cutting
- Grinding
- Edge preparation
- Rolling and L-seam fit up (Figure 2)
- L-seam welding (Figures 3 and 4)
- Back-chipping
- Die-penetrant (DP) Test (Figure 3)

Figure 2: Rolling of shell.

Figure 3: L-seam welding & DP test.

Figure 4: Joining of shell.

Figure 5: Tubes & Baffles assembly.

Channel Shell to Dish End & Flange Fit-Up

- Material identification
- Edge preparation
- Flange Fit-up and Welding
- Channel Shell Dish End Fit-up and Welding

Tubes & Baffles Assembly

- Orientation marking
- Insertion of tie rods
- Insertion of spacers and baffles
- Insertion of tubes in the baffles
- Insertion of tube bundle assembly in the shell

Tube-Sheet Fit-Up

- Angular positioning of tube-sheet with respect to shell
- Welding of tube-sheet to shell
- Second tube-sheet fit-up

Tube to Tube-Plate Joint

- Length adjustment
- Tacking of the tubes
- Dye penetrant test
- Expansion of the tubes

Channels to Main Shell Fit-Up

- Gasket Positioning
- Bolting

HYDRAULIC TESTING

For Shell Side

- Design pressure (P)
- Minimum stress ratio $(S_h/S)_{min}$
- Hydraulic test pressure at the top (P_{ht})

  $= 1.1 \times P \times (S_h/S)_{min}$

  $= 5.72$ bar

For Channel Side

- Design pressure (P)
- Minimum stress ratio $(S_h/S)_{min} = 1$
- Hydraulic test pressure at the top (P_{ht})

  $$= 1.3 \times P \times \left(S_h/S\right)_{min}$$

  $$= 6.76 \ \text{bar}$$

CONCLUSIONS

The design of STHE i.e. thermal and mechanical design was carried out using TEMA/ASME standards both manually and using software. It is found that design of STHE obtained by both approaches is very easy, simple, advance & less time consuming as comparing to existing method used in different Indian industries [such A_1 & A_2]. Manufacturing and hydraulic testing of STHE was carried out at Alfa Laval (India) ltd., Pune-12. The hydraulic test pressure at the top is found to be 6.76 bar.

REFERENCES

1. TEMA, "Standards of the Tubular Exchanger Manufacturer's Association (TEMA)," 8th Edition, Section 1-5, 7-10, Tubular Exchanger Manufacturers Association, Inc., New York, 1999.

2. ASME Section II, "For Material Specifications," 2004.

3. ASME Section V, "For Non-Destructive Examination," 2004.

4. ASME Section VIII, Division I, "Rules for Construction of Pressure Vessels," 2007.

5. ASME Section IX, "For Welding & Brazing Qualifications," 2004.

6. Wolverine, "Wolverine Tube Heat Transfer Data Book," 2nd Edition, Wolverine Tube Inc., Decatur, 1984, pp. 60-71, 85-95.

7. Dennis R. Moss, "Pressure Vessel Design Manual," 3rd Edition, Chapter 1-4, Elsevier, Amsterdam, 2003.

8. R. K. Shah, "Fundamentals of Heat Exchanger Design," 2nd Edition, Chapter 1-5, Wiley, Hoboken, 2003. doi:10.1002/9780470172605

9. R. Mukherjee, "Practical Thermal Design of Shell and Heat Exchangers," 2nd Edition, Chapter 1-5, 8, Begell House, Redding, 2004.

10. S. T. M. Than, "Heat Exchanger Design," World Academy of Science, Engineering and Technology, Vol. 46, 2008, pp. 604-611.

Chapter 5

Enhancement of Heat Transfer by Ultrasound: Review and Recent Advances

Mathieu Legay[1], Nicolas Gondrexon[1,2], Stéphane Le Person[3], Primius Boldo[4], and André Bontemps[3]

[1]LEPMI, UMR 5279, CNRS, Grenoble INP, Université de Savoie and Université Joseph Fourier BP75, 38402 Saint Martin d'Hères, France

[2]Laboratoire de Rhéologie et Procédés, UMR 5520, CNRS, Université Joseph Fourier, Grenoble I, Grenoble-INP, BP 53, 38041 Grenoble Cedex 9, France

[3]LEGI, UMR 5519, Domaine Universitaire BP 53, 38041 Grenoble Cedex 9, France

[4]EDYTEM, UMR 5204, Campus Scientifique, Université de Savoie, 73376 Le Bourget du Lac Cedex, France

ABSTRACT

This paper summarizes some applications of ultrasonic vibrations regarding heat transfer enhancement techniques. Research literature is reviewed, with special attention to examples for which ultrasonic technology was used alongside a conventional heat transfer process in order to enhance it. In several industrial applications, the use of ultrasound is often a way to increase productivity in the process itself, but also to take advantage of various subsequent phenomena. The relevant example brought forward here concerns heat exchangers, where it was found that ultrasound not only increases heat transfer rates, but might also be a solution to fouling reduction.

INTRODUCTION

In engineering applications, ultrasound is helpfully used to improve systems efficiencies. Intensifying chemical reactions, drying, welding, and cleaning are among the various possible applications of ultrasonic waves [1]. An analogous observation can be made for heat transfer processes, which are omnipresent in the industry: cooling applications, heat exchangers, temperature control, and so forth. It is somewhat logical and natural to wonder what could be the influence of ultrasound upon heat transfer systems. Strangely, it has not been a research topic deeply investigated until recently.

It appears that researches undertaken in the past concerned basic systems, usually with a single fluid, such as heating rods or walls in a volume of water subjected to ultrasonic vibrations. The tendency goes toward systems getting more complicated (e.g., cooling of tiny components, vibrating structures for heat exchangers) and models becoming more accurate with powerful numerical simulations for example.

The objectives of this paper are to provide scientific and historical backgrounds to the future studies concerning heat transfer enhancement by ultrasonic vibrations and to bring forward the evolution of this domain with several examples of applications. The first part describes an overview of ultrasound, induced phenomena, and how they positively influence heat transfer processes. Then, examples drawn from various fields of interest are analysed (thermal engineering, food industry, experimental and numerical simulations). Emphasis is made on the best improvements and results obtained. Finally, recent adaptation of ultrasonic technologies to heat exchanger devices is discussed thoroughly, with examples drawn from new patents and current laboratory work.

GENERALITIES ABOUT ULTRASOUND

Standard Classification by Power, Frequency, and Use

Acoustic waves of which frequencies are higher than the upper limit of the human hearing range, usually around 16 or 20 kHz, are called ultrasound. These waves are often classified according to their frequency or power.

Between 20 and about 100 kHz, waves are defined as "low frequency ultrasound" or "power ultrasound". Indeed, it is usually transferred at a high power level (a few tens of Watts), and therefore, ultrasound is able to modify the medium where it propagates. Power ultrasound can disrupt a fluid bulk to create cavitation or acoustic streaming, two phenomena with powerful macroscopic effects for heat transfer enhancement. Therefore, power ultrasound finds uses in various processes like cleaning, plastic welding, sonochemistry [1], and so forth. It is also generally used for heat and mass transfer processes intensification.

Further in the frequency spectrum, above 1 MHz, is found "low power ultrasound" (usually less than 10 W), at a "very high frequency" which does not affect the medium of propagation. Consequently, it is especially used for medical diagnosis or nondestructive material control, and references regarding heat transfer enhancement are very scarce in the literature.

In the intermediate range 100 kHz–1 MHz, "high frequency ultrasound" is found. It is less used than power ultrasound to promote heat transfer. Figure 1 shows some typical uses of ultrasound according to frequency and power. Thorough description of the development of ultrasonic technologies can be found in the literature [1, 2].

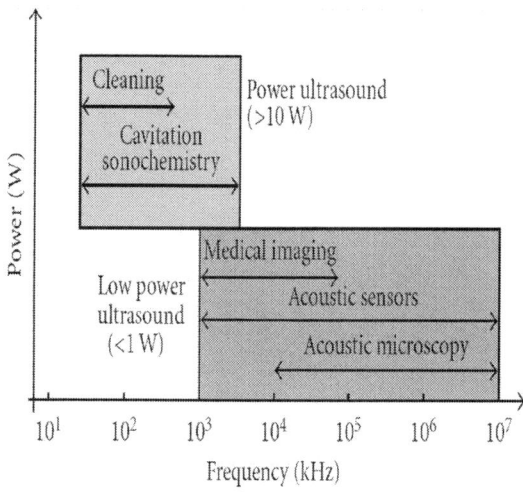

Figure 1: Utilizations of ultrasound according to frequency and power.

Ultrasound Propagation and Induced Effects

Many phenomena may ensue from propagation of an ultrasonic wave into a fluid and more particularly into a liquid medium. Two of them, of major importance for heat transfer enhancement, are acoustic cavitation and acoustic streaming. There exist other subsequent effects such as heating of the medium due to dissipation of the mechanical energy. This phenomenon is used for the determination of the ultrasonic energy supplied to the medium in an ultrasonic reactor, well-known as the calorimetric method

[1]. With high-frequency ultrasound, an acoustic fountain at the liquid-gas interface may also appear. Temperatures up to 250°C can be reached precisely at this interface [3]. Laborde et al. [4] provided a general description and mathematical modelling of some phenomena resulting from propagation of ultrasound into a liquid. Figure 2 illustrates some of these important effects that may occur in a liquid.

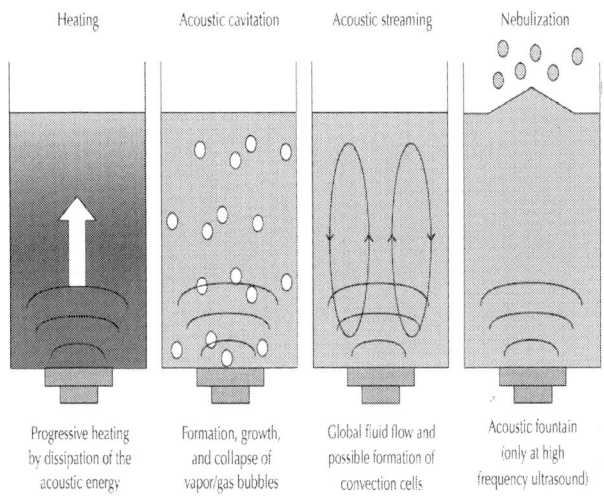

Heating	Acoustic cavitation	Acoustic streaming	Nebulization
Progressive heating by dissipation of the acoustic energy	Formation, growth, and collapse of vapor/gas bubbles	Global fluid flow and possible formation of convection cells	Acoustic fountain (only at high frequency ultrasound)

Figure 2: Four effects resulting from ultrasound propagation in a liquid.

These phenomena have always been a subject of interest since their discovery, and even though research is still ongoing, some comprehensive descriptions have been made by several authors and are frequently updated [1,4]. Therefore, this paper focuses only on two significant phenomena: acoustic streaming and acoustic cavitation, tackled from a heat transfer point of view.

Acoustic Streaming

Acoustic streaming can be considered as a well-known phenomenon since its comprehensive mathematical description by Lighthill in 1978 [5]. He explained that acoustic streaming ensues from the dissipation of acoustic energy which permits the gradients in momentum, and thereby the fluid currents. Riley [6] also makes the distinction between the quartz wind streaming happening in the fluid bulk, and the Rayleigh streaming located at the boundary layers and solid-liquid interfaces. The speed gained by the fluid allows a better convection heat transfer coefficient near the solid boundaries, sometimes leading to turbulence and promoting heat transfer rate (Figure 3).

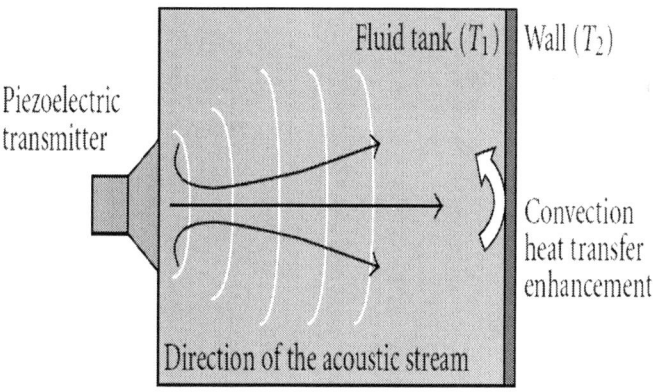

Figure 3: Acoustic streaming—enhancement of convection heat transfer.

Fand and Kave [7] foresaw in 1960 the possible effect of acoustic streaming on heat transfer intensification and studied what was named "thermoacoustic streaming", a stronger flow phenomenon than isothermal acoustic streaming.

Acoustic streaming (forced air current) was created in the air above a vibrating beam [8, 9]. It was sufficient to levitate small objects and make them spin around themselves, and thereby computing the flow velocity. The temperature of the object above the beam was decreased sensitively, and the convection heat transfer coefficient around it was increased proportionally to the stream velocity. This is an interesting first example of how acoustic streaming can modify heat transfer coefficients.

Acoustic streaming is also a factor that reduces the melting time of paraffin [10]. Its influence was studied apart and described as analogous to forced convection, whatever the profile of the standing waves field is. Nakagawa [11] even managed to simulate and control a streaming flow caused by 4 vibrators, allowing the selection of a zone that needs to be cooled down by the acoustic jet.

A type of configuration often studied is heat transfer occurring in a channel made by two plates or beams at different temperatures with vibrations applied either to the fluid between or to one of the walls [12–14].

The typical order of magnitude of acoustic streaming velocity is usually a few centimetres per second (between 1 and 100 cm s^{-1}) [9, 15], but it also appears to vary slightly with ultrasonic power and frequency [16].

Acoustic Cavitation

Acoustic cavitation is the major phenomenon that may arise from the propagation of ultrasonic waves into a liquid. Many authors have described cavitation process thoroughly but not always appearing in an oscillating pressure field, in which particular case is called acoustic cavitation [17, 18]. It is the formation, growth, oscillations, and powerful collapse of gas bubbles into a liquid. When defining acoustic cavitation, one must also describe precisely the experimental conditions at which it occurs (gas dissolution, temperature, pressure, etc.), because it depends on several parameters. When the local pressure is decreased sufficiently below the vapour pressure during the rarefaction period of the sound wave, the static pressure and the cohesive forces are overcome and gas bubbles are formed. They will generally oscillate, grow, and then collapse violently [19, 20].

There are many other ways to create cavitation into a liquid, for instance, hydrodynamic cavitation using micro-channels which can also promote cooling heat transfer [21]. Comprehensive details about acoustic cavitation in pure water can be found in [22].

There exist two types of acoustic cavitation: stable and transient [18, 23, 24]. When bubbles oscillate around an equilibrium size, this is called stable cavitation. When they exist for less than one cycle, they are transient cavities. Another important fact is that the implosion of a vaporous cavity is more violent than a gas-filled one because when vapour is turned into liquid, there is no residual gas to cushion the collapse of the bubble. Some experimental results and photographic studies showed that the impact of a collapsing cavitation bubble could last 10^{-7} s, reaching a local pressure up to 193 MPa [23]. This explains many phenomena involved in chemistry, biology, engineering, [25] and so forth. It also explains why acoustic cavitation is believed to be the major effect of ultrasonic heat transfer enhancement. Indeed, a bubble implosion near a solid-liquid interface disrupts thermal and velocity boundary layers, reducing thermal resistance and creating microturbulence, as schematically explained in Figure 4.

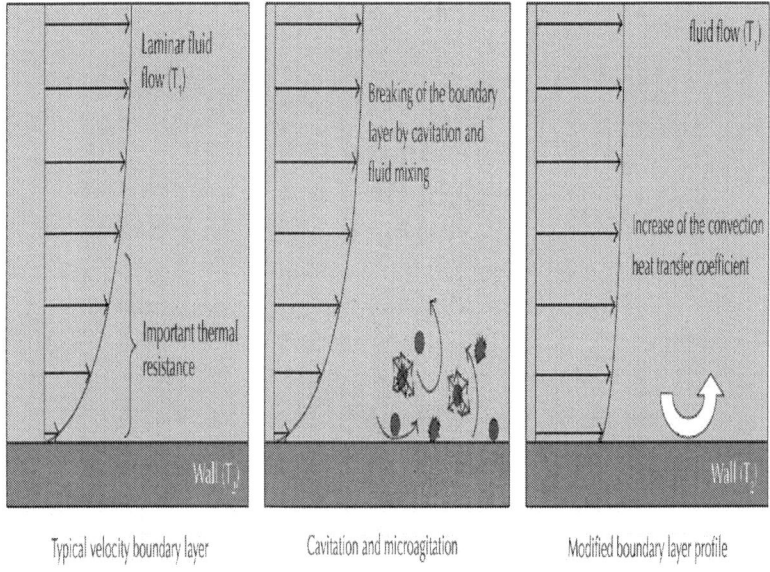

Figure 4: Explanation of heat transfer enhancement by acoustic cavitation.

Usually the bubble implosion is assumed to be of the order of the microsecond, and the bubble size is about 10^{-4} m (but also depending on frequency) [1]. So, the order of magnitude of particles displacement velocity during bubble implosion can be estimated at about $100\,\mathrm{m\,s^{-1}}$. There are approximately between 2 or 3 orders of magnitude between the acoustic streaming and the microturbulence velocities. This is one of the reasons why acoustic cavitation is often considered as the main reason for heat transfer enhancement by ultrasound. It can also be used as a way to promote or control turbulence, which already suggests some possible use in heat exchange devices. Flow friction near the boundaries could be reduced [26].

INFLUENCE OF ULTRASOUND ON HEAT TRANSFER

History

It is necessary to go back to the 60s to find the first reported studies dealing with heat transfer intensification involving ultrasonic vibrations. These very pioneer studies (see also Section 3.3.1) often gave interesting results but unfortunately, not promising enough to lead to deeper enquiries. Completely different techniques have probably been developed at the meantime (e.g., channel size reduction). Therefore, the subject was quite forgotten until the 90s, where it regained interest with the growing tendency to make more and more efficient devices for energy management. The graph proposed in Figure 5 shows the number of publications dealing with heat transfer enhancement using ultrasound, found in bibliographic databases such as Scopus and Google Scholar for 10-year periods since 1960. References taken into account are those reported in all the tables of this document. Earlier than this date, they are hard to find even if some may exist.

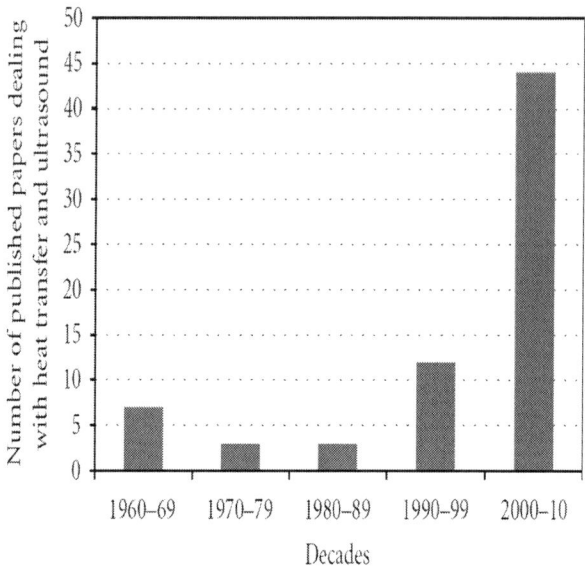

Figure 5: Evolution of the number of published papers per decade dealing with heat transfer enhancement by ultrasound.

Very few works were published in the 70s–80s but an important increase has taken place since the 90s. According to this tendency, one can expect that in the forthcoming years, this subject is likely to know a substantial development.

Among the three heat transfer modes, conduction and radiation assisted by ultrasound are the less studied. Strangely, only a few authors have investigated them although promising results were already reported in 1979 by Fairbanks [27]. He found that the combination of radiation (artificial or natural) and ultrasound to heat a flowing liquid led to better results than the sum of each process taken separately; besides, metal conduction could be enhanced between 2.25 and 5.55 times. Conversely, during melting of paraffin, when conduction was dominating over convection, Oh et al. found little influence of ultrasound [10]. This difference may be due to the nature of the materials (paraffin and metals) that have a completely different response to the vibrations. Nomura and Nakagawa [15] studied heat transfer enhancement with cavitation and acoustic streaming on a narrow surface where conduction had also a great importance. To quantify cavitation intensity, they measured the mass loss rate of a 15 μm thick aluminium foil. Microjets induced by cavitation increased the apparent thermal conductivity but they were so powerful that erosion would be a problem (e.g., for microelectronic components cooling). On a very narrow surface, conduction was always dominating over convection.

Heat Transfer with Phase Change

Melting and Solidification

Power ultrasound is a method to reduce the size of ice crystals on the frozen products and gain in quality [28]. This leads to finest ice crystals and shortens the time between the onset of crystallization and the complete formation of ice, mainly due to acoustic cavitation. Birth of nucleation sites, microstreaming, and some cleaning action of heat exchangers are among the subsequent advantages. Besides, ultrasound is a nonintrusive technique. Comprehensive reviews of the uses of ultrasound in food technology exist [29, 30], with many examples of processing, crystallization, and freezing.

The freezing temperature of supercooled water can also be controlled by ultrasonic vibrations to make ice slurry, a solid-liquid mixture very interesting to store and transport cold thermal energy. The probability of phase change is increased with the total number of cavitation bubbles, acting as nuclei for solidification inception [31, 32]. Conversely, to store warm thermal energy, ultrasound allows a melting time reduction (e.g., to take advantage of the sunlight period) without excessive electricity consumption [10]. Table 1 sums up some references where ultrasound was used for phase change applications.

Table 1: Various uses of ultrasound to promote phase change heat transfer

Reference	Description of the study	Frequency, power, intensity	Best and/or interesting result obtained
Fairbanks [27]	Radiation (Sun and infrared) into water, conduction into metal, melting hetero-geneous system	50 kHz, 61 W (radia-tion); 20 kHz, 250 W (melting); 20 kHz, 75 W (conduction)	Radiation: double heat transfer rate, conduction: 3.55 times thermal con-ductivity, melting rate doubled
Inada et al. [31]	Experimental, phase change from super-cooled water to ice, acoustic cavitation, pure water and tap water	28 kHz, 0–6.5 kW m^{-2}	Important decrease of supercooling with ultrasound for both types of water
Oh et al. [10]	Melting of paraffin in a tank with constant heat flux, acoustic streaming, cavitation, experimental and modelling study	40 kHz, 70–340 W	Melting time 72 min with 340 W ultrasonic power instead of 275 min without ultrasound
Zhang et al. [32]	Experimental study, probability of water phase change with number of bubble nuclei, cavitation, square vessel, trans-ducer at the bottom	39 kHz, 4.4 kW m^{-2}	Probability of phase change increased with number of bubble nuclei and pres-sure amplitude

Boiling

Boiling heat transfer in the presence of an ultrasonic field is described apart for being a very active research field. Ultrasound allows improvement of boiling heat transfer almost systematically. The first bubbles appearing in the nucleation sites are swept away by the vibrations, and the apparition of film boiling is therefore delayed so that higher heat fluxes are reached (see Figure 6).

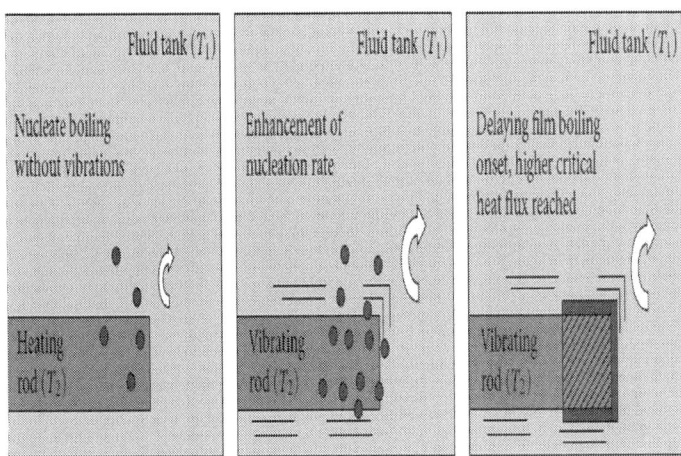

Figure 6: Enhancement of boiling heat transfer by ultrasonic vibrations.

According to several authors, this is still due to acoustic cavitation, which helps the creation and growth of the bubbles, whereas their oscillations enable to create micro-streaming and local agitation near the surfaces to sweep them away [20, 33–36]. But part of this explanation was called into question [37] because heat transfer was not enhanced at saturated liquid temperature as it should have been.

Heat transfer enhancement of saturated pool nucleate boiling was studied using a combined method: ultrasonic vibrations and glass beads (49 µm mean diameter, 120 ppm) mixed into distilled water [38]. The convection heat transfer coefficient was found up to 4 times greater.

It was reported several times that the distributions of the sound pressure and of the local heat transfer enhancement were in phase [39–43]. The critical heat flux of subcooled boiling in water in the presence of ultrasound is influenced by several parameters [44, 45]. The effect of plate inclination is reported and the optimum parameters are a surface facing the incident acoustic wave, an elevated ultrasonic power delivered and a low subcooling temperature. The critical heat flux enhancement was closely related to bubble departure from the surface, either by acoustic streaming or by microstreaming caused by cavitation. In [46], the same observation about water subcooling was made (increase of critical heat flux when subcooling temperature decreases) but a different one for the plate inclination. Park and Bergles [47] found very small increases in burnout heat flux compared to the literature with only 10 and 5%, respectively, for saturated and subcooled pool boiling. Vibrations, though not ultrasonic but induced by the flow, also allow a shifting of the critical heat flux, which strengthens the results obtained with ultrasonic vibrations [48, 49].

Table 2 summarizes some studies concerning boiling heat transfer enhancement with ultrasonic vibrations.

Table 2: Summary table of boiling heat transfer studies

Reference	Description of the study	Frequency, power, intensity	Best and/or interesting result obtained
Baffigi and Bartoli [45]	Experimental, subcooled boiling, horizontal cylinder, cavitation	40 kHz, 300–500 W	h/ h_{US}~2331/5000 W m^{-2} K^{-1} subcooling temperature: 41 K
Bergles and Newell [50]	Horizontal annulus, subcooled boiling, CHF	70 kHz; 80 kHz, 1.4 W/cm²	70 kHz, 40% local increase in non-boiling
Bonekamp and Bier [51]	Pool boiling, pure fluids (R23, R134a), and mixtures of both	42.0 kHz; 69.2 kHz; 84.7 kHz, 4 W	42 kHz, equimolar mixture, P_{US}> 1 W, 90% increase in h + important hysteresis reduction
Heffington and Glezer [36]	Pool boiling enhancement, VIBE mechanism (vibration-induced bubble ejection)	1.65 MHz	Water/ethanol ~70/30: 425% increase in CHF (600 W cm^{-2})
Jeong and Kwon [44]	CHF augmentation pool and subcooled boiling, inclination angle	40 kHz	87–126% CHF increase for downward facing surface

Kim et al. [33]	Experimental results, natural convection, pool subcooled and saturated boiling, platinum wire, transducer at the bottom, liquid FC-72	48 kHz		At least 60% global heat transfer increase (natural convection)
Kim and Jeong [52]	Numerical study, water bath, transducer at the bottom, inclination and subcooled boiling	40 kHz		see Jeong and Kwon [44]
Kwon et al. [46]	CHF enhancement pool boiling, variation of inclination angle and pool temperature, transducer at the bottom	40 kHz		CHF increased by 110% at pool temperature 95°C, horizontal downward plate
Park and Bergles [47]	Inert, dielectric liquid typical of those used for immersion cooling of microelectronic components (R-113) to cool small diameters stainless steel tubes power supplied	55 kHz, 75 W, 8000 W m^{-2}		Saturated pool: 10% increase in burnout heat flux; subcooled pool: 5% increase
Serizawa et al. [37]	Horizontal and vertical surfaces in water and vertical round tube under forced circulation of water. Silver rod at 750–800 K into distilled water (film boiling), ultrasound at the bottom	28 kHz, 70 W		Natural convection and pool nucleate boiling augmented for higher liquid subcooling. Temperature change periodically with ultrasonic waves. Quenching time reduced
Wong and Chon [20]	Natural convection and boiling around platinum wire in distilled water and methanol, cavitation, experimental work	20 kHz; 44 kHz; 108 kHz; 306 kHz, 0–200 W (with amplifier)		8-fold increase in heat transfer coefficient in natural convection

Yamashiro et al. [42,43]	Quenching process, horizontal platinum wires in subcooled water	24 kHz; 44 kHz, 0–280 W	Cooling rate and heat flux increase with cavitation intensity and water subcooling, better effect at 24 kHz
Zhou and Liu [35]	Experimental study, acetone boiling in cubic pool around an horizontal circular tube, acoustic cavitation	?	Heat transfer increased with water subcooling and cavitation intensity
Zhou [53]	Experimental investigations, copper nanofluid, acoustic cavitation, cubic vessel filled with acetone, horizontal copper tube	?	Heat transfer in presence of acoustic field increased with nanoparticles concentration, cavitation intensity, fluid subcooling
Zhou and Liu [54]	Experimental investigations, calcium-carbonate nanoparticles in acetone, acoustic cavitation, cubic vessel with horizontal copper tube	?	Convection and boiling reduced by addition of nanoparticles, but increase with acoustic field intensity
Zhou et al.[34]	Acetone boiling around horizontal copper tube in a cubic vessel, acoustic cavitation effect on boiling heat transfer	?	Higher heat flux at lower wall temperature with acoustic cavitation

Food Industry/Drying

For being particularly adequate (nonintrusive, nonchemical, etc.), ultrasonic technologies are intensively developing in food industry. Food drying is one of the best examples. If there is a good acoustic match between the transducer and the food material, ultrasonic vibrations can be directly applied to the material to be dried [55, 56]. This can produce a sponge effect, as illustrated by Figure 7: contraction and expansion cycles, leading to a better drying result. The effect is much more pronounced for very porous products, as explained in [57], which is why the porosity of the product to be dried is an important parameter to take into account before applying ultrasonic waves.

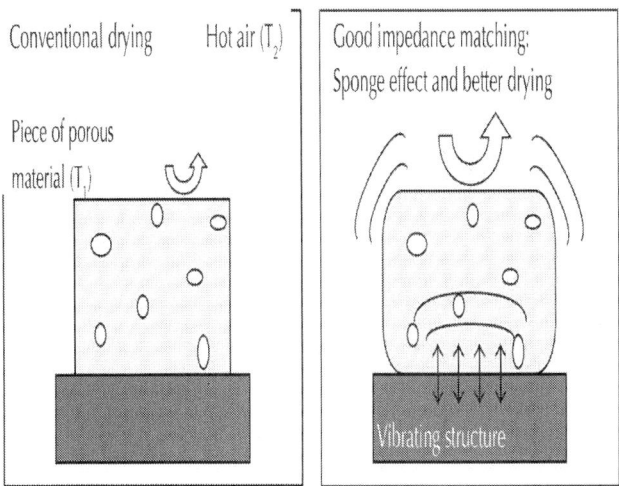

Figure 7: Sponge effect during vibration and drying of a porous food product.

Power ultrasound mainly affects the external thermal resistance. If the transducer is not in contact with the material and ultrasound is air-borne, it is reported that high air flow rate may introduce modifications in the acoustic field, decreasing also the acoustic energy transmitted to the medium. Power ultrasound increases the effective moisture diffusivities at low air velocities but the improvement becomes negligible at high air velocities [58]. A prototype of an ultrasonic dehydration system has been built and studied in [59]. An impedance matching unit was added to the vibrator to be in direct contact with the food. Applying a sufficiently high acoustic intensity, this technology would permit to save thermal energy in drying processes and to preserve the food quality.

With the aim of sterilizing food, the influence of particle size and power input on heat transfer between fluid and food size particles was investigated [60]. These parameters had little influence since the convection heat transfer coefficient was already approximately doubled every time in the presence of ultrasound. Comprehensive review of the uses of ultrasonic technology in the food industry can be found in the literature [29, 30]. A summary of some studies regarding the use of ultrasound in food industry can be found in Table 3.

Table 3: Reported uses of ultrasound in food industry

Reference	Description of the study	Frequency, power, intensity	Best and/or interesting result obtained
Cárcel et al.[58]	Drying persimmon cylinders, air velocity change, experiments, and mathematical model	21.8 kHz, 75 W, 154.3 dB	Drying speed increased with ultrasound at low air velocities (<4 m s^{-1}), affecting internal and external thermal resistances
de la Fuente-Blanco et al. [59]	Drying process with direct contact, vibrating plate	20 kHz, 0–100 W	At 100 W power, after 60 min, sample mass 27% instead of 85%
Gallego-Juárez et al. [55]	Drying process with direct contact, vibrating plate	20 kHz, 100 W	Final moisture less than 1%, speed increase, and better quality product
Li and Sun [28]	Experimental study: potatoes samples freezing into 50/50% mixture water/ethylene glycol at about −18°C	25 kHz, 7.34 W; 15.85 W; 25.89 W	Most efficient power: 15.85 W; exposure time: 2 min; during the phase change period
Mason et al. [30]	Review article (food technology)		
Sastry et al. [60]	Sterilization applications but food particles replaced by metal samples. Effect of size and power input	Power input: 0.139, 0.069 and 0.046 W g^{-1} of liquid	Convection coefficient approximately doubled in all cases
Zheng and Sun [29]	Review article (food freezing process)		

Intensification of Convection

Convection, like boiling, is one of the most studied modes of heat transfer under the influence of ultrasonic vibrations. Increases in heat transfer coefficients up to 25 times are reported [61]. Some years ago, a negligible influence of ultrasonic waves on heat transfer had been described [48, 62, 63]. But more recently, interest in this way of intensification is regained and some authors began to analyze the influence of properties of the environment of propagation (gas dissolution, temperature, flow, etc.) and characteristics of the wave itself (amplitude, frequency, etc.) [37, 38, 64]. Others examined geometries to discover new possible uses [65–67], or as in [68], studied the effect of vibrations (not ultrasonic) on the transition to turbulence and buckling flow theory. Researches undertaken in this field are summarized thereafter. When dealing with convection, it is crucial to observe that ultrasound can be considered as an "external help" to heat transfer. Therefore, it is interesting to wonder if it is not more appropriate to speak of forced convection rather than free convection when ultrasound is turned on. This fundamental question still remains opened to discussion.

Pioneer Studies

Fand and Kave [7] are among the pioneers who expected heat transfer enhancement from acoustic streaming forced convection (see Section 2.2.1). Bergles and Newell [50] were probably the first to investigate an annulus-type structure, that is, water flowing between two concentric pipes, with a heating system located inside the central pipe. In this work, up to 40% local increase of the heat transfer coefficient was reported but only 10% in the global coefficient, which was not enough for being profitable. This was in part due to the attenuation of the sound effect, or to a bad contact between the emitter and the tube containing water. Bergles [63] made a survey on the techniques to enhance heat transfer with ultrasonic vibrations. He reported that authors generally found significant increases in nonboiling heat transfer at moderate flow velocity. Improvements were clearly related to cavitation, reported not to be as effective as established boiling. The main restriction came from the attenuation of the ultrasonic energy by the vapour and the difficulties to locate the transducer so as to obtain good coupling with the fluid and suffer minimum attenuation, also reported in [50].

Conversely, in Larson's Ph.D. dissertation [62], natural and forced convection flows over a sphere were investigated. Ultrasonic frequencies, Nusselt and Reynolds numbers were the main variables. Larson claimed that cavitation was responsible for the increase in Nusselt number at the low frequencies, whereas acoustic streaming was the major factor of enhancement at higher frequencies. But he finally reported that no sufficient increase in heat transfer was obtained to warrant the use of ultrasound as a means of heat transfer intensification technique (for Reynolds numbers and ultrasonic intensities tested). Richardson [69] studied the effect of horizontal and vertical acoustic waves (710 and 1470 Hz, not ultrasound) on heat transfer around a horizontal cylinder. He found some local changes in the boundary layer thickness and consequently in convection heat transfer coefficients at high intensity sound levels.

Experiments and numerical results reported by Gould [70] showed that the heat transfer rate increased approximately linearly with the sound amplitude when water was used. Values were increased up to 10-fold with acoustic streaming. When more viscous liquids were used, the relationship between heat flow and sonic amplitude was found to be nonlinear.

Influence of Environmental and Wave Characteristics

Using frequencies below 20 kHz, Komarov and Hirasawa [64] investigated the cooling of a preheated platinum wire. Like in [8], the most efficient effect was obtained using high-amplitude sound waves. Besides, a moderate wire temperature was also necessary, otherwise cooling radiation effect was greater and convection effect diminished. This observation joins the one made in [71], where a better efficiency of ultrasonic waves at low heat fluxes is due to a thinner thermal boundary layer, easier to be disrupted by cavitation bubbles.

At a local scale, in a stationary acoustic field, it was observed that the convection heat transfer coefficient was the highest where the sound pressure was maximal [39–41]. This is due to the effect of buoyancy force coupled to pressure force and to the thermal boundary layer thickness shrinking because of water movement near the surface.

Dissolved gas can also have an influence as illustrated in [72] with gaseous cavitation into CO_2 saturated water. The distinction was between the two types of acoustic cavitation: a low-intensity gaseous cavitation, and a high-intensity vaporous cavitation. Gaseous cavitation was found to be a very good way to enhance heat transfer by increasing turbulence, in a flow where the Reynolds number is not already high. A fluid flow may also be controlled without contact (only by ultrasonic vibrations) [73]. A velocity reduction near the antinodes of the pressure wave was caused by cavitation bubbles. This effect was negligible if the flow velocity was too high because bubbles were carried downstream.

The influence of the fluid characteristics has also some importance, as shown in [74] where convective heat transfer enhancement by ultrasound was analysed into acetone, ethanol, and water. The best improvement ratio obtained was about 4-fold for acetone. Conversely, the effect of cavitation seemed different for water, brine, and sugar-water [71]. But the most probable reason for heat transfer enhancement still remains the disturbance created by cavitation bubbles and the impingement due to their implosion at the surface, causing a local thinning of the thermal boundary layer.

More unusual studies have also been undertaken like the influence of nanoparticles combined with acoustic cavitation on convection and boiling [53, 54]. Another example can be found in [75], dealing with heat transfer between a molten metal (1520°C) flowing in a tube and water around to cool it down. Convection coefficients were found to be almost doubled in the presence of ultrasound at 20 kHz.

Two graphs have been plotted in Figures 8 and 9 to sum up, respectively, the influence of the ultrasonic power supplied and the wave frequency on the increase of the convection heat transfer coefficient. Each point represents the best result obtained in the corresponding referenced paper.

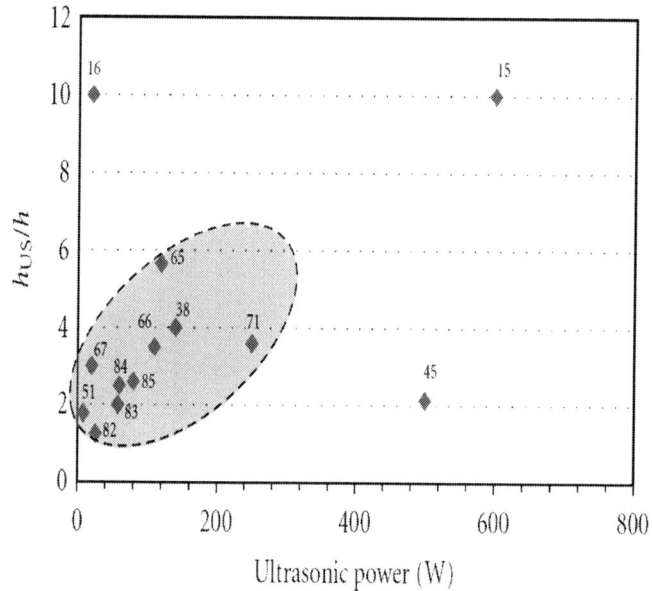

Figure 8: Increase in convection heat transfer coefficient versus ultrasonic power.

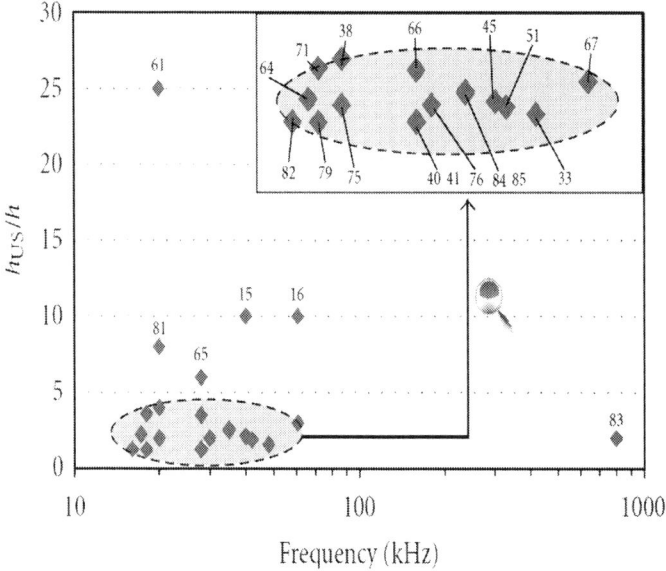

Figure 9: Influence of frequency on the increase of convection heat transfer coefficient.

One can see in Figure 8 that the intensification of convection seems to be proportional to the ultrasonic power supplied, at least for low values (<200 W, the blue zone). It would have been interesting to plot h_{US}/h as a function of the acoustic intensity (W m^{-2}) or even of the volumetric power (W m^{-3}). Unfortunately, this information is not always put forward in papers, and it is often impossible or very difficult to calculate it precisely afterwards. That is why in the future, it would be interesting and necessary to find a common term to compare studies between them (as it is already possible with frequency). However, for the moment it can be assumed that the sizes of most systems investigated in the literature are at the laboratory scale (few dozens of centimetres in length). So the plotting as a function of the total power can give a first good approximation. By the way, the 3 points outside the blue zone on Figure 8 probably correspond to references where the acoustic intensity and/or the volumetric power were very different from those in the blue zone.

Concerning Figure 9 and the effect of frequency, it is more difficult to find a tendency. Most works reported are concentrated in a zone between 15 and 60 kHz (low frequency, power ultrasound), but the improvements do not seem to depend on the frequency. More important is probably another parameter such as the system configuration or the ultrasonic power relative to surface or volume. An important point to underline is that frequencies between 60 kHz and 800 kHz (high frequency ultrasound) have not been investigated. Such frequencies would probably bring new interesting results.

Influence of the Geometry of the System

An interesting experimental setup is described in [65] to examine the effect of irradiation angle of ultrasonic waves upon the convective heat transfer rate from an inclined flat plate to water. The plate was oriented downward in front of the transducer and was electrically heated. The effect of angle of inclination on heat transfer coefficient was very low if acoustic cavitation was not generated, apart from if the plate was vertical where a small effect of acoustic streaming was detected.

Nomura et al. [66] measured experimentally the heat transfer coefficients during natural convection on a downward facing horizontal surface and a vertical surface. This coefficient was periodically changed by the ultrasonic vibrations according to the distance from the oscillator, but could become more uniform when using different ultrasonic frequencies. It also increased with the wave amplitude, as reported in [8, 70]. The distance between the transducer and the device to cool has a great importance [76]. It must be a multiple of the half wavelength used to create a resonating medium, in order to obtain more elevated acoustic streaming velocities and higher heat transfer coefficients. It is also apparently possible to create acoustic streaming behind a wall, for instance to cool the internal components of a system from the outside [67]. In [16], a horn-type transducer produced vibrations to study cooling techniques by natural convection in tap and degassed water. A convection coefficient up to 10 times higher with ultrasonic vibrations than without was calculated and different regions where the enhancement was more or less pronounced were observed. An interesting and original use of power ultrasound is for wood treatment [77]. Ultrasonic waves could have a very positive effect on the temperature increase speed in the centre of wood cylinders which are either air-dried or fully water-saturated.

Sum Up of Convection Studies

In the domain of ultrasonically improved heat transfer, convection is the most studied area, as illustrated by Figure 10.

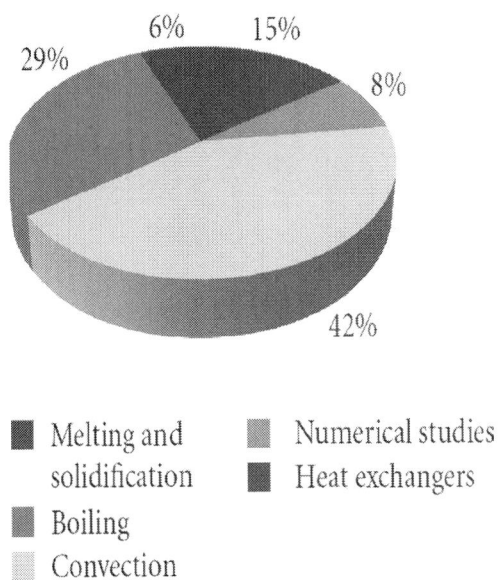

Figure 10: Percentage of studies by subject (total: 62 papers from the tables of this document).

This chart was made with all references quoted in the tables except Table 3 (food) because many other studies exist in this domain and it would not have been representative (e.g., see [29, 30]). Convection covers at least half of these studies, and even more because it appears also in heat exchangers and in phase change.

A very important point is the cause of heat transfer enhancement, which is very difficult to determine since many phenomena appear simultaneously during propagation of ultrasound. Figure 11 shows a diagram with these different phenomena and the number of times these effects are assumed to be the cause of heat (and mass) transfer enhancement. This diagram was made from references of all the tables of this text (except the reviews), but more than one effect can be quoted in one paper (which explains why the number of effects quoted is superior to the number of papers).

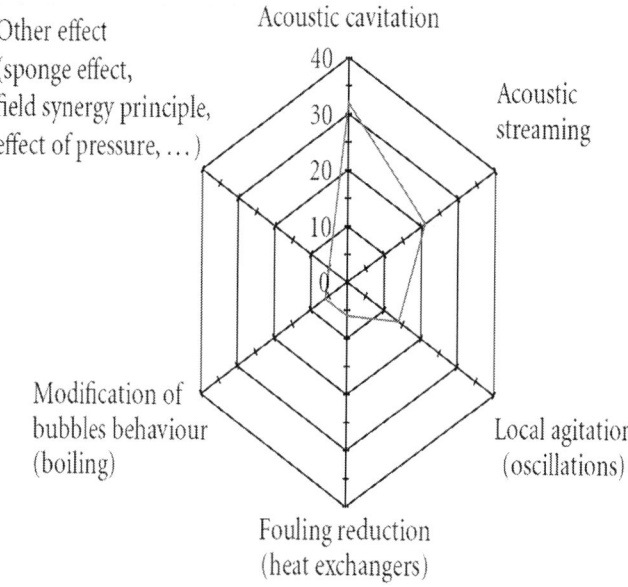

Figure 11: Ultrasound effects held responsible for heat transfer enhancement.

According to this statistic chart, acoustic cavitation is the predominant phenomena for heat transfer enhancement. It is followed by acoustic streaming and by local agitation due to oscillations. Other phenomena, such as fouling reduction, hysteresis reduction, change in bubble behaviour, are side effects that could become very important when ultrasound will be used in industrial systems.

Table 4 synthesizes improvements obtained for convection heat transfer assisted by ultrasonic waves.

Table 4: Ultrasonic waves and convection heat transfer improvements

Reference	Description of the study	Frequency, power, intensity	Best and/or interesting result obtained
Bergles [63]	Review article, heat transfer enhancement		
Cai et al. [71]	Experimental, natural convection, acoustic cavitation, circular heated copper tube in water, brine and sugar water	18 kHz, 0–250 W	Heat flux from cylinder: 132 W m^{-2}, ultrasonic intensity: 80 W cm^{-2}, enhancement up to 360%.
Fand and Kave [7]	Acoustic streaming, convection heat transfer, heated cylinder	800 Hz–4800 Hz (no ultrasound)	3-fold increase in heat transfer rate
Gould [70]	Acoustic streaming, convection between metal and water or glycerin-water mixtures	?	Up to 10-fold increase
Hoshino and Yukawa [41]	Experimental investigation, hot and cold cylinders, vertical standing waves, local and global coefficients in degassed water	28 kHz, 0.1–0.215 W cm^{-2}	Local coefficient $h_{US}/h \approx 1.3$ at 0.125 W cm^{-2}, maximum at antinode and minimum at node
Hoshino et al. [40]	Free convective heat transfer from a heated wire	28 kHz	Local coefficient $h_{US}/h \approx 1.25$ at 0.24 W cm^{-2} acoustic intensity, maximum at antinode and minimum at node

Hyun et al. [8]	Experiments and CFD simulations of acoustic streaming induced by flexural vibrations of a beam, cooling of a stationary beam above	28.4 kHz	Temperature drop of 40°C in 4 min, h up to 157 W m^{-2} K^{-1}
Iida et al. [39]	Experimental, natural convection heat transfer from a fine cylinder to water, comparison convection coefficient and sound pressure profiles	28 kHz	Augmentation ratio around 1.6 when $\Delta P >$ 0.02 MPa
Komarov and Hirasawa [64]	Standing and travelling sound waves in tubes, platinum wire	0.3–17.2 kHz	$Nu_{US} / h \approx 2.25$ at 17.2 kHz, no gas flow and preheated wire temperature ~675 K
Lam et al. [77]	Experimental study, saturated and air-dried wood cylinders heated in a water bath at 59.8°C with and without ultrasound Temperature recorded at the centre of the cylinders	50–55 kHz, commercial cleaner	Significant influence of ultrasound on the temperature increase at the centre of cylinders
Larson [62]	Acoustic streaming around a sphere within a cylinder, cavitation, toluene, and water	20–1000 kHz, up to 6 W cm^{-2}	Increase in Nusselt number up to about 4 times, but not sufficient to warrant the technology
Lee and Loh [76]	Acoustic streaming in a gap between heat source and transducer	30 kHz	Heat transfer rate increased up to 75%
Lee and Choi [72]	Acoustic cavitation into CO_2 saturated water	138 W	Up to 369.5% turbulence intensity enhancement

Loh et al. [9]	Experiments and simulations (CFX4), flexural vibrations of a beam, acoustic streaming in air above (open) to cool a fixed beam	28.4 kHz	Temperature drop of 40°C in 4 min, streaming velocity up to 2 m s^{-1}
Markov et al. [75]	Flowing molten metal (~1520°C) in a water-cooled tube	20 kHz	Heat transfer coefficient as much as doubled
Nakagawa [11]	Experimental and computational results (CFX4), 4 vibrators to control acoustic streaming in a vessel containing silicon oil	1 MHz	Maximum streaming velocity measured: 0.07 m s^{-1}, jet position modified
Nakayama and Kano [38]	Experiments, cylindrical glass vessel, distilled water, with or without glass beads	20 kHz, 0–140 W	With glass beads, at saturation pressure 13.3 kPa, increased up to 4 times
Nomura and Nakagawa [15]	Cooling a narrow surface, acoustic streaming and cavitation effects separated, water tank, experimental investigations	40 kHz, 600 W	Acoustic streaming at 0.4 m s^{-1}, predicted with forced convection equations. Cavitation: increased up to about 10 times
Nomura et al. [16]	Downward facing surface, ultrasound from below, experimental, cavitation, and acoustic streaming	60.7 kHz, 5–20 W	Up to 10-fold increase in heat transfer coefficient, tap and degassed water
Nomura et al. [26]	Turbulence intensity measured experimentally, square channel, transducer at the bottom	25 kHz, 0–50 W	Turbulence intensity 3 times larger with ultrasonic vibrations and up to 5 times locally

Nomura et al. [66]	Effect of ultra-sonic frequency on downward facing and vertical surface	28 kHz (110 W), 45 kHz (210 W), 100 kHz (25 W)	Around 2 or 3 times average increase in
Nomura et al. [67]	Experimental, natural convec-tion, obstacle in front of a heating surface (different materials), acous-tic streaming	60.7 kHz, 5–20 W	Up to 3 times with acrylic plate at 20 W, obstruction plates placed near the horn tip
Richardson [69]	Horizontal heated cylinder, horizon-tal and vertical sound fields, shadowgraphs	710 and 1470 Hz (no ultrasound), 120–140 dB	Local changes of boundary layer thick-ness and heat transfer enhancement
Uhlenwinkel et al. [61]	Experimental, gas vessel (air argon helium), resonant acoustic field, distance between transducers 20–200 mm	10 and 20 kHz	Heat transfer enhance-ment up to 25 times at ambient pressure at about 0.9 MPa and 20 kHz
Vainshtein et al. [12]	Two horizontal plates at differ-ent temperatures, acoustic streaming in longitudinal direction	200 Hz–15 kHz, 140 and 145 dB	Nu from 1 to 10, in-crease with frequency
Yukawa et al. [65]	Inclined copper plate in water	28 kHz, 0.1–0.48 W cm^{-2}	Convection coefficient increased 6-fold at in-clination 90°, intensity 0.48 W cm^{-2}
Zhou et al. [74]	Horizontal copper tube in water, ac-etone and ethanol, experimental study	?	Maximum ratio of heat transfer enhancement: 3.95 with acetone, maximum source inten-sity, and close sound distance

Numerical Studies, Modelling

Numerical simulation is taking a more and more important place with the growing potential of computational calculation. Even if the systems of interest often remain quite simple (one fluid, one moving part), the level of accuracy of computations can be very high [8, 9, 11, 52]. At least four equations have to be solved when dealing with numerical problems involving heat transfer and acoustic waves: continuity, momentum (Navier-Stokes), energy, and a least one for the streaming forces (from Nyborg's theory [78]). If acoustic cavitation is modelled, equations must be solved for the two fluid phases (liquid, vapour). Vibration is usually represented by a moving boundary and a dynamic mesh modelling (e.g., [52]) or by a sound field distribution inside the liquid (e.g., [79]).

A numerical model of acoustic streaming between two parallel beams separated by an air gap between 0.1 and 2 mm wide is proposed in [80]. The Nusselt number is increased only by 1% under constant heat flux conditions and by 0.5% under heat source condition. The initial purpose was to study the feasibility of cooling computer chips in laptops. The 2D simulation described in [81] showed that a standing waves pattern was necessary to obtain an increase in heat transfer. The reason for heat transfer enhancement invoked in [52] is fluid mixing by ultrasonic vibrations that provided fresh water to the heat transfer surface, increasing the temperature gradient. Wave and flow patterns can be predicted precisely, which could be a basis of a future tool for the optimization of vibrating heat exchangers [13].

The field synergy principle is also a convincing way to illustrate cavitation-enhanced heat transfer [79]. This principle says that the local temperature gradient vectors should be parallel to the local velocity vectors to obtain the better convection heat transfer effect. And indeed, it is seen on streamlines patterns and temperature gradient patterns that acoustic cavitation helps to reduce in many zones the intersection angle between these two vector fields. Table 5 summarizes enhancements observed by undertaking numerical simulations.

Table 5: Summary of numerical researches on convection increase by ultrasound

Reference	Description of the study	Frequency, power, intensity	Best and/or interesting result obtained
Aktas et al.[81]	Shallow enclosure, vibrating vertical side wall, acoustic streaming	20 kHz and 25 kHz	**After 5 ms,** $h / h_{US} / h \approx 40 / 320$ **at 20 kHz, $\Delta T = 10$ K**
Cai et al. [79]	Square enclosure—hot bottom, natural convection, acoustic cavitation, ultrasonic beam from the centre	18 kHz	Field synergy principle analysis, 25% increase in h at the centre
Lin and Farouk [13]	Gas-filled square enclosure vibrating side-wall, top-side heated	20 kHz	Heat transfer enhanced with streaming flow velocity (maximum at the middle of the bottom wall)
Wan and Kuznetsov [80]	Acoustic streaming in a gap (0.1– 2 mm) between two horizontal beams, the lower vibrating	160 Hz (no ultrasound)	1% increase in Nusselt number for constant heat flux case of the upper beam
Wan and Kuznetsov [14]	Air channel composed of two parallel beams, upper beam vibrating	21 kHz	h from 0.9 to 82 W m^{-2} K^{-1} at constant heat flux, decreasing with channel width

APPLICATIONS TO HEAT EXCHANGERS

In the previous sections, examples concerned configurations with only one fluid in thermal contact with another solid body at a different temperature. It was necessary to gain a good knowledge of those basic systems before studying more complex ones. Heat exchangers have at least two fluids (flowing or at rest), which makes systems sometimes more tricky to study. Indeed, they are subjected to several constraints, and ultrasonic vibrations have influence on various parameters (e.g., heat transfer, fouling, and charge losses). It is, therefore, more difficult to assess the efficiency of ultrasound on such systems. That is probably one of the main reasons why their development is quite recent. This is the field of research that is currently under development in our laboratory.

Examples from the Literature

One of the first studies was carried out by Kurbanov and Melkumov in 2003 [82]. They explained comprehensively why ultrasonic vibrations are very well suited to increase performances of liquid-to-liquid heat exchangers. According to them, acoustic waves homogenize the velocity vectors of the subflows in pipes and decrease the surface tension of the fluid near the boundaries. The latter phenomenon is even more interesting if a thin film of lubricant is stuck to the pipes surfaces, which usually happens in refrigeration systems. This thin film induces a thermal resistance and its removal is very interesting for performances improvement.

Cooling of sonochemical reactors by cold water flowing into a coil, as presented in Figure 12, was experimentally and analytically analysed [83]. The cooling time of a certain amount of water, stored in the chemical reactor, was compared with and without high-frequency ultrasonic vibrations. The convection coefficient was enhanced between 135 and 204% in the presence of acoustic waves, reducing effectively the cooling time. Observed improvement was explained in terms of overall agitation due to the combined effects of local mixing (acoustic cavitation) and global fluid motion within the reactor (acoustic streaming).

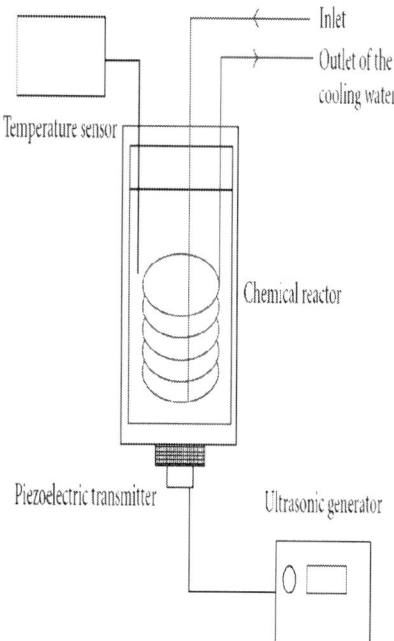

Figure 12: Ultrasonically assisted cooling of a chemical reactor.

A shell-and-tube configuration for a fluid-to-fluid vibrating heat exchanger was built and studied [84, 85]. This system is presented in Figure 13.

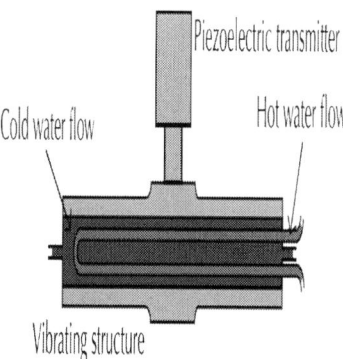

Figure 13: Schematic diagram of the vibrating shell and tube heat exchanger.

The ratio between the overall heat transfer coefficient with ultrasound and the one without ultrasound for this shell-and-tube heat exchanger was calculated and found ranging from 1.2 up to 2.6 depending on the liquid flowrate at the shell side [85]. The ultrasonic power had negligible influence on the heat exchange rate and the overall heat transfer coefficient was always higher with ultrasound than without, whatever the liquid flow rates or range of temperatures tested. Further investigations on the same system showed that higher improvements could be expected, especially for slow laminar flows in the shell.

Other Subsequent Advantages

As shown in [51], ultrasonic vibrations could be interesting to achieve a complete activation of nucleation sites in large evaporators with extended surfaces, normally reached with a sufficiently high heat flux (and consequently elevated wall temperature). Indeed, ultrasound is efficient to reduce hysteresis effect [86], that is, the tendency of a system to remain in its initial state in spite of the cause supposed to produce a change. Another important phenomenon resulting from ultrasonic vibrations application and not described until here is surface cleaning (essentially thanks to acoustic cavitation). This is very important because it could be part of a solution to reduce the natural fouling process in heat exchangers. Indeed, the environmental conditions in such devices make them prone to corrosion or microorganisms deposition. They induce additional thermal resistances which prevent the system from operating in optimal conditions, adding environmental and economical costs. However, one must pay attention to the powerful erosion capability of cavitation that could damage materials. Benzinger et al. [87] have studied the effect of ultrasound on a microstructured heat exchanger to avoid fouling. Their results are very promising because the convection heat transfer coefficient increases almost up to the initial value after an ultrasonic pulsation cycle. Biofouling control is a possible application of ultrasound, that is, the prevention of microorganism growth (algae, fungi, bacteria) on surfaces by application of ultrasonic vibrations [88]. Other examples are the synergistic properties of axially propagated ultrasound and antibiotic on the removal of biofilms in water-filled tubes [89]. An analogous study analysed the combined effect of ozone and ultrasonic vibrations [90]. The result observed was that the use of ozone and ultrasound was more effective than each process alone. But optimal parameters are sometimes difficult to find, for example, concerning scale removal [91] with choice of temperature, distance, and acoustic intensity.

Influence of ultrasound on pressure drop, or charge losses, also seems to be positive although very few studies deal with this subject [26, 82, 92]. Table 6 sums up the different examples of vibrating heat exchangers and of positive effects of ultrasound on these systems encountered in the research literature.

Table 6: Review of vibrating structures for heat exchangers and their advantages

Reference	Description of the study	Frequency, power, intensity	Best and/or interesting result obtained
Benzinger et al. [87]	Microstructured heat exchanger, antifouling investigations	20 kHz, 35 W	Pulses of 1 min to break the fouling layer but fouling speed increased
Bott and Tianqing [90]	Ozone + ultrasound to clean heat exchangers, axially propagated ultrasound	20 kHz, 2357.8 kW m^{-2}	2357.8 kW m^{-2}, 3 × 1 min pulse/day, up to 70% reduction in biofilm thickness
Bott [88]	Control of biofilm formation or biofilm removing in heat exchangers	20 kHz	88% reduction of biofilm growth with 10 treatments/day, 3 × 30 s at 40% amplitude
Gondrexon et al. [85]	Vibrating shell-and-tube heat exchanger, experimental investigation	35 kHz, 80 W	Overall heat transfer coefficient increased up to 257%
Kurbanov and Melkumov [82]	Heat exchanger-type for heating and refrigeration	3 and 16 kHz	27% increase in hbut other major advantages

Li et al. [91]	Effects of various parameters on antiscale and scale removal. Sedimentary speed and scale inhibition rate analysed	14–20 kHz; 0–250 W	Larger acoustic intensity is better for scale removal. 40°C best for antiscale, 50°C for scale removal. Better effect for small distances to the ultrasonic transducer
Monnot et al. [83]	Cooling of chemical reactor (2.9 L), experimental and modelling	800 kHz; 1.6 MHz; 20 kHz; 0–109 W	Max $h_{US} / h \approx 2.04$ ~ 2.04 at 800 kHz, 57.6 W
Mott et al. [89]	Experimental investigation, glass tubes filled with water, standing waves	20–350 kHz, 35–45 W	95.3% of biofilm removed by 2 × 30 s treatment at 20 kHz in 7 cm tubes, 87.5% at 3 × 30 s in 50 cm tubes
Tisseau et al. [84]	Shell and tube heat exchanger, experimental investigation	35 kHz, variable power	Overall heat transfer coefficient increase up to 250%

Patented Devices

Assessments of all these advantages in academic research literature are rare. Nevertheless, several systems (setups) regarding vibrating heat exchangers have been recently patented [93–104]. Almost all of them claim energy consumption savings either by a fouling reduction (or cleaning effectiveness) [93–97] or an improvement of the heat exchange efficiency, and sometimes both of them. These patents may involve different types of structures such as shell and tube heat exchangers [94, 95, 98, 99], water tank and heating coil (batch configuration like in Figure 12) [100], or various heat exchanger devices with applications in chemical engineering (reducing reaction time [101], increasing defrosting speed [102], cryogenic applications [103], and steelmaking applications [104]).

CONCLUSIONS

Ultrasound has gained a growing interest from industry during the last decades, resulting in the development of several specific applications. Ultrasonic waves appear as an interesting way to improve processes productivity especially to overcome transfer limitations. For what concerns heat transfer, ultrasound can also be regarded as a possible technical solution for heat exchange enhancement. Hence, a lot of publications dealing with fundamental studies can be found in the literature. But most of these works are performed at the laboratory scale involving academic set-ups and usually using classical low frequency ultrasound. Well-known ultrasonicallyinduced effects such as acoustic cavitation, acoustic streaming, and fluid particles oscillations are responsible for heat transfer improvement observed. It is also very important to note here that it is very difficult to distinguish the influence of these effects since they often occur simultaneously. One might therefore consider the positive influence of ultrasound as an overall effect. As detailed in this paper, influence of ultrasound on convection remains the major subject of interest. Local heat transfer coefficient was shown to be multiplied between 2 and 5 times in the presence of an ultrasonic field. Phase change heat transfer also covers a great number of studies that demonstrate the beneficial effect of ultrasound on boiling as well as melting or solidification. A more recent and scarce research field that focuses on heat exchangers has shown that the use of ultrasonic waves can improve overall performances regarding heat transfer and/or fouling.

Although very promising results are reported, the scale-up of the ultrasonic technology to pilot or industrial scale heat exchangers has not been yet deeply investigated. Only few references are available in the literature, illustrating the difficulties to meet such a technological challenge. It is then expected that the combined efforts of acousticians, chemical and mechanical engineers will also help to design a new type of "vibrating" heat exchangers. It might, therefore, result in improved performances as well as antifouling action in the near future.

ACKNOWLEDGMENTS

This work was supported by the association Instituts Carnot, Energies du futur.

REFERENCES

1. T. J. Mason and J. P. Lorimer, Sonochemistry: Theory, Applications and Uses of Ultrasound in Chemistry, Ellis Horwood, Chichester, UK, 1988.

2. J. D. N. Cheeke, Fundamentals and Applications of Ultrasonic Waves, CRC Press, New York, NY, USA, 2002.

3. H. Li, Y. Li, and Z. Li, "The heating phenomenon produced by an ultrasonic fountain," Ultrasonics Sonochemistry, vol. 4, no. 2, pp. 217–218, 1997.

4. J. L. Laborde, A. Hita, J. P. Caltagirone, and A. Gerard, "Fluid dynamics phenomena induced by power ultrasounds," Ultrasonics, vol. 38, no. 1, pp. 297–300, 2000. ·

5. S. J. Lighthill, "Acoustic streaming," Journal of Sound and Vibration, vol. 61, no. 3, pp. 391–418, 1978. ·

6. N. Riley, "Acoustic streaming," Theoretical and Computational Fluid Dynamics, vol. 10, no. 1–4, pp. 349–356, 1998.

7. R. M. Fand and J. Kave, "Acoustic streaming near a heated cylinder," Journal of the Acoustical Society of America, vol. 32, pp. 579–584, 1960.

8. S. Hyun, D. R. Lee, and B. G. Loh, "Investigation of convective heat transfer augmentation using acoustic streaming generated by ultrasonic vibrations," International Journal of Heat and Mass Transfer, vol. 48, no. 3-4, pp. 703–718, 2005.

9. B. G. Loh, S. Hyun, P. I. Ro, and C. Kleinstreuer, "Acoustic streaming induced by ultrasonic flexural vibrations and associated enhancement of convective heat transfer," Journal of the Acoustical Society of America, vol. 111, no. 2, pp. 875–883, 2002.

10. Y. K. Oh, S. H. Park, and Y. I. Cho, "A study of the effect of ultrasonic vibrations on phase-change heat transfer," International Journal of Heat and Mass Transfer, vol. 45, no. 23, pp. 4631–4641, 2002.

11. M. Nakagawa, "Analyses of acoustic streaming generated by four ultrasonic vibrators in a vessel,"Japanese Journal of Applied Physics, vol. 43, no. 5, pp. 2847–2851, 2004.

12. P. Vainshtein, M. Fichman, and C. Gutfinger, "Acoustic enhancement of heat transfer between two parallel plates," International Journal of Heat and Mass Transfer, vol. 38, no. 10, pp. 1893–1899, 1995. ·

13. Y. Lin and B. Farouk, "Heat transfer in a rectangular chamber with differentially heated horizontal walls: effects of a vibrating sidewall," International Journal of Heat and Mass Transfer, vol. 51, no. 11-12, pp. 3179–3189, 2008.

14. Q. Wan and A. V. Kuznetsov, "Streaming in a channel bounded by an ultrasonically oscillating beam and its cooling efficiency," Numerical Heat Transfer A, vol. 45, no. 1, pp. 21–47, 2004.

15. S. Nomura and M. Nakagawa, "Ultrasonic enhancement of heat transfer on narrow surface," Heat Transfer, vol. 22, no. 6, pp. 546–558, 1993.

16. S. Nomura, A. Yamamoto, and K. Murakami, "Ultrasonic heat transfer enhancement using a horn-type transducer," Japanese Journal of Applied Physics, vol. 41, no. 5, pp. 3217–3222, 2002.

17. R. E. Apfel, "Acoustic cavitation inception," Ultrasonics, vol. 22, no. 4, pp. 167–173, 1984.

18. E. A. Neppiras, "Acoustic cavitation series: part one. Acoustic cavitation: an introduction," Ultrasonics, vol. 22, no. 1, pp. 25–28, 1984.

19. S. Nomura, K. Murakami, and Y. Sasaki, "Streaming induced by ultrasonic vibration in a water vessel,"Japanese Journal of Applied Physics, vol. 39, no. 6, pp. 3636–3640, 2000.

20. S. W. Wong and W. Y. Chon, "Effects of ultrasonic vibrations on heat transfer to liquids by natural convection and by boiling," AIChE Journal, vol. 15, no. 2, pp. 281–288, 1969.

21. B. Schneider, A. Ko ar, C. J. Kuo et al., "Cavitation enhanced heat transfer in microchannels," Journal of Heat Transfer, vol. 128, no. 12, pp. 1293–1301, 2006.

22. E. Herbert, Cavitation acoustique dans l'eau pure, Ph.D. thesis, École normale supérieure, Département de physique, 2006.

23. E. A. Neppiras, "Acoustic cavitation," Physics Reports, vol. 61, no. 3, pp. 159–251, 1980.

24. E. A. Neppiras, "Acoustic cavitation thresholds and cyclic processes," Ultrasonics, vol. 18, no. 5, pp. 201–209, 1980.

25. E. Webster, "Cavitation," Ultrasonics, vol. 1, no. 1, pp. 39–48, 1963.

26. S. Nomura, K. Murakami, and M. Kawada, "Effects of turbulence by ultrasonic vibration on fluid flow in a rectangular channel," Japanese Journal of Applied Physics, vol. 41, no. 11, pp. 6601–6605, 2002.

27. H. V. Fairbanks, "Influence of ultrasound upon heat transfer systems," in Proceedings of the Ultrasonics Symposium, pp. 384–387, 1979.

28. B. Li and D. W. Sun, "Effect of power ultrasound on freezing rate during immersion freezing of potatoes," Journal of Food Engineering, vol. 55, no. 3, pp. 277–282, 2002.

29. L. Zheng and D. W. Sun, "Innovative applications of power ultrasound during food freezing processes—a review," Trends in Food Science and Technology, vol. 17, no. 1, pp. 16–23, 2006.·

30. T. J. Mason, L. Paniwnyk, and J. P. Lorimer, "The uses of ultrasound in food technology," Ultrasonics Sonochemistry, vol. 3, no. 3, pp. S253–S260, 1996.

31. T. Inada, X. Zhang, A. Yabe, and Y. Kozawa, "Active control of phase change from supercooled water to ice by ultrasonic vibration 1. Control of freezing temperature," International

Journal of Heat and Mass Transfer, vol. 44, no. 23, pp. 4523–4531, 2001.

32. X. Zhang, T. Inada, A. Yabe, S. Lu, and Y. Kozawa, "Active control of phase change from supercooled water to ice by ultrasonic vibration 2. Generation of ice slurries and effect of bubble nuclei,"International Journal of Heat and Mass Transfer, vol. 44, no. 23, pp. 4533–4539, 2001. ·

33. H. Y. Kim, Y. G. Kim, and B. H. Kang, "Enhancement of natural convection and pool boiling heat transfer via ultrasonic vibration," International Journal of Heat and Mass Transfer, vol. 47, no. 12-13, pp. 2831–2840, 2004.

34. D. W. Zhou, D. Y. Liu, X. G. Hu, and C. F. Ma, "Effect of acoustic cavitation on boiling heat transfer,"Experimental Thermal and Fluid Science, vol. 26, no. 8, pp. 931–938, 2002.

35. D. Zhou and D. Liu, "Boiling heat transfer in an acoustic cavitation field," Chinese Journal of Chemical Engineering, vol. 10, no. 5, pp. 625–629, 2002.

36. S. Heffington and A. Glezer, "Enhanced boiling heat transfer by submerged ultrasonic vibrations," inProceedings of the Therminic 2004, Sophia Antipolis, France, October 2004.

37. A. Serizawa, M. Mukai, N. Aoki, et al., "Effect of ultrasonic emission on boiling and non-boiling heat transfer in natural and forced circulation systems," in Proceedings of the X International Heat Transfer Conference, vol. 6, pp. 97–102, Brighton, UK, 1994.

38. A. Nakayama and M. Kano, "Enhancement of saturated nucleate pool boiling heat transfer by ultrasonic vibrations," Heat Transfer, vol. 20, no. 5, pp. 407–417, 1991.

39. Y. Iida, K. Tsutsui, R. Ishii, and Y. Yamada, "Natural-convection heat transfer in a field of ultrasonic waves and sound pressure," Journal of Chemical Engineering of Japan, vol. 24, no. 6, pp. 794–796, 1991.

40. T. Hoshino, H. Yukawa, and H. Saito, "Effect of ultrasonic vibrations on free convective heat transfer from heated wire to water," Heat Transfer, vol. 5, no. 1, pp. 37–49, 1976.

41. T. Hoshino and H. Yukawa, "Physical mechanism of heat transfer from heated and cooled cylinder to water in ultrasonic standing wave field," Journal of Chemical Engineering of Japan, vol. 12, no. 5, pp. 347–352, 1979.

42. H. Yamashiro, H. Takamatsu, and H. Honda, "Effect of ultrasonic vibration on transient boiling heat transfer during rapid quenching of a thin wire in water," Journal of Heat Transfer, vol. 120, no. 1, pp. 282–286, 1998.

43. H. Yamashiro, H. Takamatsu, and H. Honda, "Enhancement of cooling rate during rapid quenching of a thin wire by ultrasonic vibration," Heat Transfer, vol. 27, no. 1, pp. 16–30, 1998. ·

44. J. H. Jeong and Y. C. Kwon, "Effects of ultrasonic vibration on subcooled pool boiling critical heat flux,"Heat and Mass Transfer, vol. 42, no. 12, pp. 1155–1161, 2006.

45. F. Baffigi and C. Bartoli, "Heat transfer enhancement from a circular cylinder to distilled water by ultrasonic waves in subcooled boiling conditions," in Proceedings of the ITP2009 Interdisciplinary Transport Phenomena VI: Fluid, Thermal, Biological, Materials and Space Sciences, Volterra, Italy, October 2009.

46. Y. C. Kwon, J. T. Kwon, J. H. Jeong, and S. H. Lee, "Experimental study on CHF enhancement in pool boiling using ultrasonic field," Journal of Industrial and Engineering Chemistry, vol. 11, no. 5, pp. 631–637, 2005.

47. K. A. Park and A. E. Bergles, "Ultrasonic enhancement of saturated and subcooled pool boiling,"International Journal of Heat and Mass Transfer, vol. 31, no. 3, pp. 664–667, 1988.

48. Z. W. Douglas, M. K. Smith, and A. Glezer, "Acoustically enhanced boiling heat transfer," in Proceedings of the Therminic 2007, Budapest, Hungary, September 2007.

49. D. H. Kim, Y. H. Lee, and S. H. Chang, "Effects of mechanical vibration on critical heat flux in vertical annulus tube," Nuclear Engineering and Design, vol. 237, no. 9, pp. 982–987, 2007.

50. A. E. Bergles and P. H. Newell, "The influence of ultrasonic vibrations on heat transfer to water flowing in annuli," International Journal of Heat and Mass Transfer, vol. 8, no. 10, pp. 1273–1280, 1965.

51. S. Bonekamp and K. Bier, "Influence of ultrasound on pool boiling heat transfer to mixtures of the refrigerants R23 and R134A," International Journal of Refrigeration, vol. 20, no. 8, pp. 606–615, 1997.

52. H. J. Kim and J. H. Jeong, "Numerical analysis of experimental observations for heat transfer augmentation by ultrasonic vibration," Heat Transfer Engineering, vol. 27, no. 2, pp. 14–22, 2006.

53. D. W. Zhou, "Heat transfer enhancement of copper nanofluid with acoustic cavitation," International Journal of Heat and Mass Transfer, vol. 47, no. 14–16, pp. 3109–3117, 2004.

54. D. W. Zhou and D. Y. Liu, "Heat transfer characteristics of nanofluids in an acoustic cavitation field,"Heat Transfer Engineering, vol. 25, no. 6, pp. 54–61, 2004.

55. J. A. Gallego-Juárez, G. Rodriguez-Corral, J. C. Galvez-Moraleda, and T. S. Yang, "A new high-intensity ultrasonic technology for food dehydration," Drying Technology, vol. 17, no. 3, pp. 597–608, 1999.

56. J. V. García-Pérez, J. A. Cárcel, E. Riera, and A. Mulet, "Influence of the applied acoustic energy on the drying of carrots and lemon peel," Drying Technology, vol. 27, no. 2, pp. 281–287, 2009.

57. J. V. García-Pérez, J. A. Cárcel, J. Benedito, and A. Mulet, "Power ultrasound mass transfer enhancement on food drying," Food and Bioproducts Processing, vol. 85, no. 3, pp. 247–254, 2007.

58. J. A. Cárcel, J. V. García-Pérez, E. Riera, and A. Mulet, "Influence of high-intensity ultrasound on drying kinetics of persimmon," Drying Technology, vol. 25, no. 1, pp. 185–193, 2007.

59. S. de la Fuente-Blanco, E. R. F. de Sarabia, V. M. Acosta-Aparicio, A. Blanco-Blanco, and J. A. Gallego-Juárez, "Food drying process by power ultrasound," Ultrasonics, vol. 44, pp. e523–e527, 2006.

60. S. K. Sastry, G. Q. Shen, and J. L. Blaisdell, "Effect of ultrasonic vibration on fluid-to-particle convective heat transfer coefficients," Journal of Food Science, vol. 54, pp. 229–230, 1989.

61. V. Uhlenwinkel, R. Meng, and K. Bauckhage, "Investigation of heat transfer from circular cylinders in high power 10 kHz and 20 kHz acoustic resonant fields," International Journal of Thermal Sciences, vol. 39, no. 8, pp. 771–779, 2000.

62. M. B. Larson, A study of the effects of ultrasonic vibrations on convective heat transfer in liquids, Ph.D. dissertation, Stanford University, Stanford, Calif, USA, 1961, Mic 61-1235.

63. A. E. Bergles, "Survey and evaluation of techniques to augment convective heat and mass transfer,"Progress in Heat and Mass Transfer, vol. 1, pp. 331–424, 1969.

64. S. Komarov and M. Hirasawa, "Enhancement of gas phase heat transfer by acoustic field application,"Ultrasonics, vol. 41, no. 4, pp. 289–293, 2003.

65. H. Yukawa, T. Hoshino, and H. Saito, "Effect of ultrasonic vibration on free convective heat transfer from an inclined plate in water," Heat Transfer, vol. 5, no. 4, pp. 1–16, 1976. ·

66. S. Nomura, K. Murakami, Y. Aoyama, and J. Ochi, "Effects of changes in frequency of ultrasonic vibrations on heat transfer," Heat Transfer, vol. 29, no. 5, pp. 358–372, 2000.

67. S. Nomura, M. Nakagawa, S. Mukasa, H. Toyota, K. Murakami, and R. Kobayashi, "Ultrasonic heat transfer enhancement with obstacle in front of heating surface," Japanese Journal of Applied Physics, vol. 44, no. 6, pp. 4674–4677, 2005.

68. H. Engelbrecht and L. Pretorius, "The effect of sound on natural convection from a vertical flat plate,"Journal of Sound and Vibration, vol. 158, no. 2, pp. 213–218, 1992.

69. P. D. Richardson, "Local effects of horizontal and vertical sound fields on natural convection from a horizontal cylinder," Journal of Sound and Vibration, vol. 10, no. 1, pp. 32–41, 1969.

70. R. K. Gould, "Heat transfer across a solid-liquid interface in the presence of acoustic streaming," The Journal of the Acoustical Society of America, vol. 40, no. 1, pp. 219–225, 1966.

71. J. Cai, X. Huai, S. Liang, and X. Li, "Augmentation of natural convective heat transfer by acoustic cavitation," Frontiers of Energy and Power Engineering in China, vol. 4, no. 3, pp. 313–318, 2010.

72. S. Y. Lee and Y. D. Choi, "Turbulence enhancement by ultrasonically induced gaseous cavitation in the CO_2 saturated water," KSME International Journal, vol. 16, no. 2, pp. 246–254, 2002.

73. S. Nomura, Y. Sasaki, and K. Murakami, "Flow pattern in a channel during application of ultrasonic vibration," Japanese Journal of Applied Physics, vol. 39, no. 8, pp. 4987–4989, 2000.

74. D. Zhou, X. Hu, and D. Liu, "Local convective heat transfer from a horizontal tube in an acoustic cavitation field," Journal of Thermal Science, vol. 13, no. 4, pp. 338–343, 2004.

75. A. V. Markov, Y. S. Astashkin, and I. I. Sulimtsev, "Influence of ultrasound on heat transfer under the conditions of forced flow of a high-temperature melt," Journal of Engineering Physics and Thermophysics, vol. 48, no. 2, pp. 242–244, 1985.

76. D. R. Lee and B. G. Loh, "Smart cooling technology utilizing acoustic streaming," IEEE Transactions on Components and Packaging Technologies, vol. 30, no. 4, pp. 691–699, 2007.

77. F. Lam, S. Avramidis, and G. Lee, "Effect of ultrasonic vibration on convective heat transfer between water and wood cylinders," Wood and Fiber Science, vol. 24, no. 2, pp. 154–160, 1992.

78. W. L. Nyborg, "Acoustic streaming near a Boundary," Journal of Acoustical Society of America, vol. 4, pp. 329–339, 1958.

79. J. Cai, X. Huai, R. Yan, and Y. Cheng, "Numerical simulation on enhancement of natural convection heat transfer by acoustic cavitation in a square enclosure," Applied Thermal Engineering, vol. 29, no. 10, pp. 1973–1982, 2009.

80. Q. Wan and A. V. Kuznetsov, "Numerical study of the efficiency of acoustic streaming for enhancing heat transfer between two parallel beams," Flow, Turbulence and Combustion, vol. 70, no. 1–4, pp. 89–114, 2003.

81. M. K. Aktas, B. Farouk, and Y. Lin, "Heat transfer enhancement by acoustic streaming in an enclosure,"Journal of Heat Transfer, vol. 127, no. 12, pp. 1313–1321, 2005.

82. U. Kurbanov and K. Melkumov, "Use of ultrasound for intensification of heat transfer process in heat exchangers," in Proceedings of the International Congress of Refrigeration, vol. 4, pp. 1–5, Washington, DC, USA, 2003.

83. A. Monnot, P. Boldo, N. Gondrexon, and A. Bontemps, "Enhancement of cooling rate by means of high frequency ultrasound," Heat Transfer Engineering, vol. 28, no. 1, pp. 3–8, 2007.

84. Y. Tisseau, P. Boldo, N. Gondrexon, and A. Bontemps, "Conception et étude préliminaire d'un échangeur de chaleur tubes et calandre assisté par ultrasons," in Proceedings of the 18ème Congrès Français de Mécanique, Grenoble, France, Août 2007.

85. N. Gondrexon, Y. Rousselet, M. Legay, P. Boldo, S. Le Person, and A. Bontemps, "Intensification of heat transfer process: improvement of shell-and-tube heat exchanger performances by means of ultrasound,"Chemical Engineering and Processing, vol. 49, no. 9, pp. 936–942, 2010.

86. D. W. Zhou, "A novel concept for boiling heat transfer enhancement," Journal of Mechanical Engineering, vol. 51, no. 7-8, pp. 366–373, 2005.

87. W. Benzinger, U. Schygulla, M. Jäger, and K. Schubert, "Antifouling investigations with ultrasound in a microstructured heat exchanger," in Proceedings of the 6th International Conference on Heat Exchanger Fouling and Cleaning—Challenges and Opportunities, Kloster Irsee, Germany, 2005.

88. T. R. Bott, "Biofouling control with ultrasound," Heat Transfer Engineering, vol. 21, no. 3, pp. 43–49, 2000.

89. I. E. C. Mott, D. J. Stickler, W. T. Coakley, and T. R. Bott, "The removal of bacterial biofilm from water-filled tubes using axially propagated ultrasound," Journal of Applied Microbiology, vol. 84, no. 4, pp. 509–514, 1998.

90. T. R. Bott and L. Tianqing, "Ultrasound enhancement of biocide efficiency," Ultrasonics Sonochemistry, vol. 11, no. 5, pp. 323–326, 2004.

91. H. X. Li, X. L. Huai, J. Cai, and S. Q. Liang, "Experimental research on antiscale and scale removal by ultrasonic cavitation," Journal of Thermal Science, vol. 18, no. 1, pp. 65–73, 2009.

92. X. L. Duan, X. Y. Wang, G. Wang, Y. Z. Chen, and X. Q. Qiu, "Experimental study on the influence of ultrasonic vibration on heat transfer and pressure drop in heat exchanger tubes," Petrochemical Equipment, vol. 33, no. 1, pp. 1–4, 2004.

93. Z. Zhenxian, "Ultrasonic highly effective heat exchanger," CN 201364049 (Y), December 2009.

94. Z. Jili and M. Liangdong, "Shell and tube type acoustic cavitation sewerage heat exchanger," CN 201229131 (Y), April 2009.

95. Z. Jili and M. Liangdong, "Immersion type acoustic cavitation sewerage heat exchanger," CN 201229132 (Y), April 2009.

96. K. Tanaka, "Heat exchanger provided with ultrasonic vibration function," JP 2006349239 (A), December 2006.

97. T. Makino, "Heat exchanger," JP 2007120933 (A), May 2007.

98. Y. Ye, L. Shiqing, and C. Jing, "Pipe type heat exchanger with heat exchange shell intensified by ultrasonic wave," CN 101196380 (A), June 2008.

99. H. Dezhong, "Superturbulent heat exchanger cavitated and reinforced by ultrasonic wave and a heat exchange method thereof," CN 101586924 (A), November 2009.

100. H. J. Robionek, "Hot water tank heat exchanger has ultrasound directed at the centre of the heating coil, to swirl the water around it," DE 102007040031 (A1), February 2009.

101. Y. Yingwu, L. Yuanchen, X. Xiangyang, et al., "Apparatus of ultrasound heat exchange," CN 101091838 (A), December 2007.

102. C. Zhenqian, "Air source heat pump ultrasound wave defrosting system," CN 101144669 (A), March 2008.

103. K. Tanaka, "Heat exchanger with vibrator to remove accumulated solids," US 2009032222 (A1), December 2006.

104. M. Nogues, "Process and apparatus to vibrate a continuous metal casting input mold," FR 8907839 (A), June 1989.

Energy from Waste: Reuse of Compost Heat as a Source of Renewable Energy

G. Irvine, E. R. Lamont, and B. Antizar-Ladislao

School of Engineering, Institute for Infrastructure and Environment, University of Edinburgh, William Rankine Building, The King's Buildings, Edinburgh EH9 3JL, UK

ABSTRACT

An in-vessel tunnel composting facility in Scotland was used to investigate the potential for collection and reuse of compost heat as a source of renewable energy. The amount of energy offered by the compost was calculated and seasonal variations analysed. A heat exchanger was designed in order to collect and transfer the heat. This allowed heated water of 47.3°C to be obtained. The temperature could be further increased to above 60°C by passing it through multiple tunnels in series. Estimated costs for installing and running the system were calculated. In order to analyse these costs alternative solar thermal and ground source heat pump systems were also designed. The levels of supply and economic performance were then compared. A capital cost of £11,662 and operating cost of £1,039 per year were estimated, resulting in a cost of £0.50 per kWh for domestic water and £0.10 per kWh for spatial heat. Using the heat of the compost was found to provide the most reliable level of supply at a similar price to its rivals.

INTRODUCTION

Composting is an aerobic process where organic materials are biologically decomposed, producing mainly compost, carbon dioxide, water, and heat. Conventional composting processes typically comprise four major microbiological stages in relation to temperature: mesophilic, thermophilic, cooling, and maturation, during which the structure of the microbial community also changes, and the final product is compost [1]. In recent years, the development and widespread use of more expensive in-vessel systems for the processing of biowastes has resulted from legislative pressures on the safety of the composting process and the subsequent use of the compost product [2]. Such systems allow for much more precise control of the composting process particularly in terms of moisture and temperature control [3]. Thus, current composting approaches and technologies tend to emphasize the use of high temperatures (> 7 0°C) in order to meet regulatory requirements for pathogen control [2].

Compost has been widely used as soil conditioners and soil fertilizers. This practice is recommended, as soil fertility needs more than ever to be sustained. Food demand is increasing rapidly in non-OECD (Organisation for Economic Co-operation and Development) countries, and it is in those countries particularly where organic waste needs to be diverted from landfill sites to composting practices, so compost can enhance soil fertility [4]. In OECD countries, where composting of organic waste is already established, its use as a landfill cover to abate greenhouse gas emissions has shown to be promising [5]. The addition of compost can minimize land degradation and soil erosion. Additionally, composting can contribute to achieve sufficient hygienisation of organic wastes and control soilborn and airborn pathogens by promotion of beneficial micro-organisms and suppression of harmful micro-organisms [6].

As energy demand is increasing rapidly, bionergy is seen as one of the primary possibilities for preventing global warming [7]. At present, the immediate factor impeding the emergence of an industry converting biowastes into bioenergy on a large scale is the high cost of processing, rather than the cost or availability of biomass feedstock [8]. Thus, the challenge is to extend the amount of bioenergy that can be produced sustainably by using biowastes, such as municipal, industrial, and construction waste as biomass feedstocks [9]. Thus, it is suggested that the heat generated during composting processes can be reused as a renewable source of energy.

A limited number of previous studies has investigated the potential energy content of compost. A recent study reports that during high-temperature phases (~60°C) of municipal waste composting, on average $1136\,kJ\,kg^{-1}$ of heat was released [10]. Similar values ($961\,kJ\,kg^{-1}$) have been reported earlier with an average compost moisture content of 52.7% [11]. Heat produced during the composting of wheat straw and poultry droppings was approximately $17.06\,MJ\,kg^{-1}$ [12] and $12.8\,MJ\,kg^{-1}$ [13], respectively. Additionally, it has been reported that the compost from municipal waste is characterised by fairly low values of thermal conductivity coefficient ($0.31\,Wm^{-1}K^{-1}$ for a compost density of $600\,kg\,m^{-3}$, 60°C), and that an increase in temperature or density both lead to an increase in the thermal conductivity coefficient [10]. Thus, as the compost ages, and it suffers a reduction in density and temperature, the thermal conduction coefficient will decrease. Klejment and Rosi ski (2008) concluded that the low value of heat conductivity coefficients does not allow compost to cool too fast and enables the application of a battery of heat exchangers. A limited number of studies on compost heat reuse has also been reported. Lekic [14] investigated the increase in water temperature between the inlet and the outlet of polyethylene pipes embedded in composting windrows and reported that 73% of the theoretical value of heat energy was transferred to the water. One main limitation of this study was the placement of the pipes within the compost mass. A solution proposed by Seki and Komori [15] involved using a packed column heating tower that transfers the heat from the warm exhaust air of the compost to a volume of water.

From the above, it is fully justified to investigate the potential reuse of compost heat as a source of renewable energy. Our primary objectives were (a) to identify the potential energy available through full-scale in-vessel composting units; (b) to identify potential technological solutions to harnessing/collecting the energy; (c) to identify the optimum alternative use for the energy collected; and (d) to critically evaluate the potential of reusing that heat from compost by comparing the performance with alternative renewable energies.

MATERIALS AND METHODS

Site Description

This case study was focused upon the Deerdykes Composting Facility located in Scotland, UK. The facility was originally constructed on the site of a decommissioned sewage treatment works, where much of the existing infrastructure was able to be reused for the composting facility. Work was completed in 2006, and the site currently accepts green waste, industrial sludge, and liquid waste [16]. The main components of the site are the site office, the in-vessel composting tunnels, the windrow composting area, and the raw material mixing area (Figure 1).

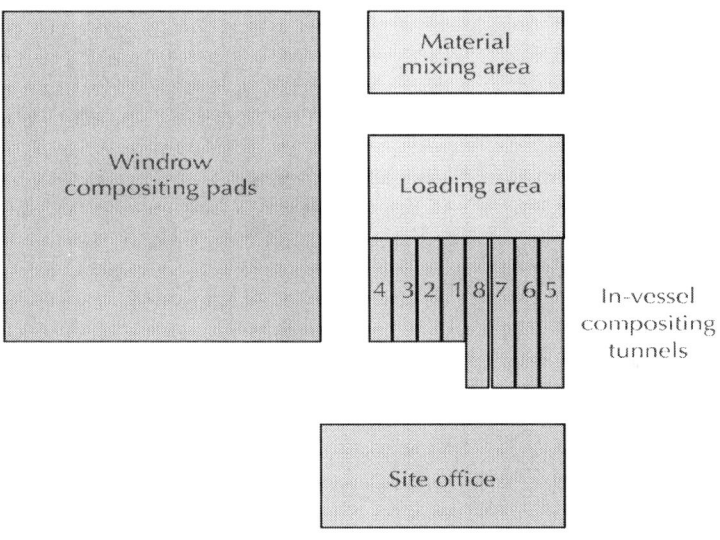

Figure 1: Deerdykes site layout.

In-Vessel Units' Description and Operation

There are 8 in-vessel composting tunnels that were constructed in the former presettlement tanks [16]. The dimensions of the tunnels varied with tunnels 1–4 being 5 m wide and 25 m long, tunnels 5 and 6 being 5.3 m wide and 35 m long, and tunnels 7 and 8 being 5 m wide and 35 m long. The tunnels were all approximately 5 m in height, however, compost was only loaded to a height of 3 m (Figure 2). Compost was loaded for an average period of 12–17 days, allowing a sanitary and stable condition of compost to be achieved. Air was supplied through small aeration holes in the floor of the tunnel thus ensuring aerobic conditions were maintained throughout the compost mass. The air drawn off from the top was primarily recirculated through the compost with a small portion expelled as exhaust air. Additionally, fresh air was mixed with the recirculated air to ensure oxygen concentrations were maintained at acceptable levels.

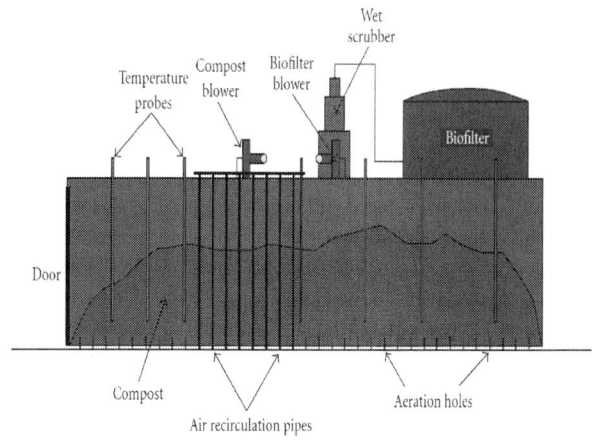

Figure 2: In-vessel unit cross-section.

The exhaust air was put through two stages of treatment. Firstly, the wet scrubber removed ammonia, hydrogen sulphide, and volatile fatty acids whilst cooling the warm air. Secondly, the biofilter removed any remaining concentrations of the pollutants and provided odour control, allowing the air to be released to the atmosphere. Potential production of methane, a potent green house gas, in anaerobic "pockets" within the composting pile, would have either metabolically oxidized to carbon dioxide while the percolated through the composting pile, or during its passage through the biofilter. Thus emissions of green house gases were not expected.

The in-vessel composting process was completely computer controlled by a software package specifically designed for the process. The air flow rates for all air blowers were varied automatically depending on the current temperature, oxygen, and pressure levels inside the tunnel. A user-defined minimum oxygen content was maintained whilst aiming to maximise the temperature, and thus degradation rate, of the compost [16]. Once the two-week period was complete the degraded compost material was transferred to the neighbouring windrow pads where it was allowed to decompose for a further 6 weeks.

Temperature Profiles

Temperature values for each batch of compost and in-vessel unit were gathered during January–December 2008 for analysis. The data stated the temperatures measured by each of the 8 temperature probes that were placed into the mass of compost inside the tunnel. Results were recorded with a regular time step of 15 minutes, with the occasional exception, throughout the whole of the degradation process. The 8 temperature values were averaged for each batch in order to derive an overall time-temperature relationship. Any readings that were clearly incorrect were disregarded.

Individual time-temperature relationships were then accumulated by season allowing overall trends to be analysed. Subsequently an average time-temperature relationship for each of the four seasons could be created.

On-Site Hot Water Demand

The average value for daily hot water consumption for the Deerdykes composting facility's site office was calculated at $180\,L\,d^{-1}$. A medium heating requirement of 12 kW for the site office was estimated, which ranged from poorly isolated buildings with a very high heating load of 22 kW [17] to an Eco-House with a required heating load of 8 kW [18]. It was assumed that the heating would be used on average 980 hours per year.

Energy Values

Energy values were then calculated for each of the four seasons. The method used was of that described by Haug [1] using the standard heat flow into a substance at constant pressure equation. This equation was also utilised in energy-balanced analyses carried out by other researchers in composting systems [19].

The following equation was used:

$$q_p = m \times c_p \times \Delta T, \tag{1}$$

where q_p is heat energy leaving the system (kJ kg^{-1}), m is mass of product (kg), c_p is specific heat at a constant pressure (kJ kg^{-1}K^{-1}), and DT is change in temperature (K). This equation assumes a process at constant pressure with a constant specific heat capacity. This assumption is valid for the relatively small changes in pressure and temperature associated with composting [1]. Heat energy values were calculated for each time step using estimated compost material concentrations and standard specific heat capacity values (Table 1). The energy values were then accumulated by day to give the energy stored in kJ kg^{-1}day^{-1}. Cumulative energy over the 15-day composting period was also calculated.

Table 1: Estimated material composition for composting at Deerdykes

Material	% present	C_p (kJ kg^{-1}K^{-1})	OverallC_p (kJ kg^{-1}K^{-1})
Air	10%	1.012	2.844
Water	60%	4.184	
Soil	28%	0.80	
Lignocellulosic material	2%	0.42	

Heat Exchange and Energy Collection

A heat exchanger was designed for the collection of energy generated during composting. The heat exchanger selected for the purpose of this study was a pipeline made of stainless steel that run suspended from the top of the in-vessel tunnels in the airspace above the composting piles. At the initial calculation stage the length and diameter of pipe required was unknown so the initial pipe layout shown in Figure 3 was investigated. The design of the heat exchanging element of the proposed design was carried out using the principles and methods discussed by Shah and Sekulic [20].

Figure 3: Initial design pipe layout.

Nomenclature used in the following section is summarized in Table 2. The aim of the design process was to determine the outlet temperature for both the hot and cold fluid for a suggested heat exchanger surface area

Table 2: Nomenclature

Parameter[a]	Description	Unit
T_i	Temperature of the fluid entering the system	K
T_o	Temperature of the fluid leaving the system	K
m	Mass flow rate of the fluid	$kg\,s^{-1}$
c_p	Specific heat capacity of the fluid at constant pressure	$kJ\,kg^{-1}K^{-1}$
d_i	Inner diameter of the heat exchanger pipe	m
d_o	Outer diameter of the heat exchanger pipe	m
k_w	Thermal conductivity value of the pipe wall material	$W\,m^{-1}K^{-1}$
h_{AIR}	Heat transfer coefficient of air	$W\,m^{-2}K^{-1}$
h_{WATER}	Heat transfer coefficient of water	$W\,m^{-2}K^{-1}$
A	Surface area of the heat exchanger wall	m^2

[a]The subscript h refers to the initially hotter fluid, and the subscript c refers to the initially colder fluid.

Hot Fluid

In this case study, the hot fluid was the warm, moist air that is released by the composting process and circulated throughout the tunnel. The temperature of this air was dictated by the temperature of the compost itself; the majority of the air was recirculated. The temperature of the air approached that of the compost after the process has been running for a certain period. The only other effect on the temperature of the air was through heat conductive losses through the concrete walls and roof of the tunnel. The mass flow rate of the air through the tunnel was driven by a centrifugal blower the maximum capacity of which was 8000 m³ hr⁻¹. These blowers were controlled in real time by computer to regulate oxygen and moisture levels.

Cold Fluid

It was proposed to use water as the cold fluid that runs through the pipe work of the heat exchanger. This was due to its low cost, ease of availability, and its thermal properties which were optimal for absorbing and storing thermal energy [21]. The mass flow rate of the water was fully controllable by the design team. A pump, pressure, or gravity fed system was designed according to requirements.

Individual Heat Transfer Coefficient

The individual heat transfer coefficient was calculated using

$$h = \frac{\Delta Q}{A \times \Delta T \times \Delta t},$$

(2)

where DQ is the heat input or heat lost (J), h is the heat transfer coefficient, (W m^{-2}K^{-1}), A is the heat transfer surface area (m^2), DT is the temperature difference between the solid surface and surrounding fluid (K), and Dt is the time period (s).

Overall Heat Transfer Coefficient

The overall heat transfer of the pipe was calculated by summing the individual heat transfer coefficients of the acting fluids using [20]

$$\frac{1}{U} = \frac{1}{h_o} + \frac{1}{h_{o,f}} + \frac{d_o \ln(d_o/d_i)}{2k_w} + \frac{d_o}{h_{i,f}d_i} + \frac{d_o}{h_i d_i},$$

(3)

where the subscripts o and i refer to the outside and inside of the pipe wall, respectively, U is the overall heat transfer coefficient (W m^{-2}K^{-1}), and h$_{r_f}$ is the thermal fouling resistance capacity (W m^{-2}K^{-1}).

Design Equations

The heat capacity rates for each fluid were calculated using [20]

$$C = \dot{m}\, c_p.$$

(4)

Subsequently the heat capacity ratio, C*, could be calculated using (5). The heat capacity ratio is simply the smaller-to-larger ratio of the heat capacity rates of the two fluids

$$C^* = \frac{C_{min}}{C_{max}}.$$

(5)

The next step was to calculate the ratio of the overall thermal conductance to the smaller of the two heat capacities, which is defined as the number of transferred units (NTUs). This was found with the following equation:

$$NTU = \frac{UA}{C_{min}}.$$

(6)

Once these values were obtained the exchanger effectiveness, ε, was calculated. The equation for exchanger effectiveness depends on the type and flow direction associated with the particular exchanger being designed. In this case study (7) was used as it is appropriate for the counter-flow conditions that existed in the tunnel [20]

$$\varepsilon = \frac{1 - \exp[-NTU(1 - C^*)]}{1 - C^* \exp[-NTU(1 - C^*)]}. \tag{7}$$

Once the exchanger effectiveness was calculated the fluid outlet temperature was found using

$$\varepsilon = \frac{C_h(T_{h,i} - T_{h,o})}{C_{min}(T_{h,i} - T_{c,i})} = \frac{C_c(T_{c,o} - T_{c,i})}{C_{min}(T_{h,i} - T_{c,i})}. \tag{8}$$

Comparison of Waste to Energy with Other Renewable Energies

Compost heat as a source of renewable heat was compared to solar thermal systems and to ground-source heat. The solar thermal system was designed using the good-practice guidelines discussed by DGS [17], which was originally written with a single-family house in mind which proved transferable to the purpose in this case study. Design methods were discussed separately for both domestic hot water supply and spatial heating.

Supplying domestic hot water is the most common use for solar thermal systems. The following sizing calculations allowed a full design to be proposed. Using the calculated value of hot water demand for the site office, $V_{H,W}$, the heat requirement was determined using

$$Q_{HW} = V_{HW} \times c_p \times \Delta T, \tag{9}$$

where $Q_{H,W}$ is the daily heat requirement (kWh day^{-1}), $V_{H,W}$ is the daily hot water consumption (L day^{-1}), c_p is the specific heat capacity of water (Wh kg^{-1}K^{-1}), and DT is the temperature difference between the hot and cold water (K).

In order to calculate the area of the solar collector required the desired solar fraction, SF, and the overall average system efficiency, η_{SYS}, of the solar collector were found. The SF is the ratio of solar heat yield to total energy required by the building and is shown by (10). It showed what percentage of the yearly heat energy demand is to be supplied by solar rather than conventional means

$$SF = \frac{Q_S}{Q_S + Q_{AUX}} \times 100,$$

(10)

where SF is the desired solar fraction, Q_s is the solar heating requirement (kWh), and Q_{AUX} is the auxiliary heating requirement (kWh).

Achieving as high a solar fraction as possible would appear desirable, however due to the variable nature of solar energy throughout the year in temperate zones it is advisable to aim for a solar fraction of around 60% [17]. Aiming for a solar fraction of 60% prevented the supply of hot water becoming overly stressed during the winter months, due to the provision of a backup boiler supply.

If aiming to counter this by using a large area of solar collectors to better cope with winter months, it will result in an oversupply of hot water during the summer months, thus a much less efficient design. For these reasons as the solar fraction increases the system efficiency decreases. When coupled with the high set-up costs associated with a scheme of that kind it proved to be an option that limits the economic attractiveness of solar thermal systems [17].

The average system efficiency is the ratio of solar heat yield to global solar irradiance experienced by the absorber surface and is linked to the solar fraction. Average system efficiencies η_{SYS} take into account losses at the collector, solar circuit, and storage. Guidelines state that initial calculations should assume a η_{SYS} of 0.35 for a flat-plate collector and 0.45 for an evacuated-tube collector [17]. This data was then used to calculate the required area of solar collector using (11). The yearly solar irradiance value, E_g was calculated for Glasgow (Table 3):

$$A = \frac{365 \text{ days} \times Q_{HW} \times SF}{E_G \times \eta_{SYS}},$$

(11)

where A is absorber surface area (m²), $Q_{H,W}$ is the daily heat re-quirement (kWh day^{-1}), SF is the desired solar fraction, E_G is the yearly potential solar irradiance (kWh m^{-2} year^{-1}), and η_{SYS} is the average system efficiency.

Table 3: Monthly solar irradiation (kW h m^{-2}) for Glasgow, UK (DGS 2005)

Month	Jan	Feb	Mar	Apr	May	Jun	Jul	Aug	Sep	Oct	Nov	Dec
Irradiance	0.45	1.04	1.94	3.40	4.48	4.70	4.35	3.48	2.33	1.26	0.60	0.32

In order to calculate the optimal diameter for the piping of the solar circuit it is vital to regulate both the speed of flow and the volumetric flow. In order to minimise noise nuisance and prevent abrasion a flow speed, , of 0.7 m s^{-1} is to be aimed for [17]. The level of volumetric flow is key in keeping the collector cooling at an efficient rate, preventing overheating and therefore wasting energy. It has been shown that a volumetric flow of about 40 L hr^{-1} per m² of collector area is ideal [17]. The volumetric flow was calculated using

$$\dot{m} = \frac{\dot{Q}}{c_p \times \Delta T},$$

(12)

where $\dot{m}$ is the volumetric flow (L m^{-2}hr^{-1}), $\dot{Q}$ is the usable thermal output converted by the collector (W m^{-2}hr^{-1}), c_p is the specific heat capacity of the solar fluid (kJ kg^{-1}K^{-1}), and DT is the temperature difference between the feed and the return flows (K).

Subsequently the optimum pipe diameter, D, could be calculated using

$$D = \sqrt{\frac{4(\dot{m}/v)}{\pi}}.$$

(13)

This calculation allowed an appropriate size of commercially available pipe to be proposed. The recommended method for calculating the collector area required to fulfill spatial heating demand is currently far less developed than for domestic hot water supply [17]. This is due to the highly variable nature of the thermal insulation of buildings, individual preferences for the comfortable temperature of a room, and whether the building uses conventional or underfloor heating systems. The calculation method used was of that described by DGS (2005) which is based solely on the living area required to be heated. The relationships presented in Table 4 are valid for a climate with a low solar fraction of 35% such as the UK and were used to calculate the collector surface area and storage volume that would be required [17].

Table 4: Design guidelines for solar thermal spatial heating system

Parameter	Recommended value
Evacuated tube collector area	0.5–0.8 m^2 of collector area per 10 m^2 heated living area
Storage volume	50 L per m^2 of collector surface area

Regarding a ground source heat pump design, generic design guidelines for specifying ground source heat pumps are currently at an underdeveloped stage.

RESULTS AND DISCUSSION

Temperature and Energy Values

Figure 4 shows the seasonal average temperature temporal profiles between January and December 2008. The compost reached temperatures above 60°C after two days, and the highest values (~70°C) were recorded during the summer. This time is similar (2–4 days) to that observed when composting with one direction of airflow [22], but higher than that observed (0.63 days) using high recirculation of processed air [19]. Average temperatures at the end of day 12 of composting were above 65°C during spring, summer, and autumn, and about 60°C during the winter, which was close to the values reported by Harper et al. [23] for 1.5–2 m of compost depth and 7 days of composting and by Ekinci et al. [19] for pilot scale 208 L reactors, 0.285 m in radius, and 0.816 m in height.

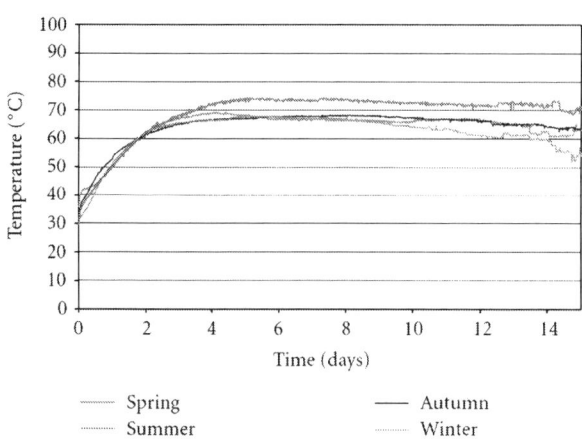

Figure 4: Average temperature temporal profiles: seasonal comparison.

Energy values were calculated using these average temperature profiles, and they are presented in Figure 5, where Figure 5(a) presents daily energy values and Figure 5(b) presents cumulative energy values. In this study, maximum heat generation rate per initial mass of compost dry matter was ~6000 kJ kg^{-1} day^{-1}. This value was higher than that reported by Harper et al. [23] for straw and poultry manure composting (2791 kJ kg^{-1} day^{-1}), and by Ekinci et al. [19] for paper mill sludge with broiler litter (2435 kJ kg^{-1} day^{-1}). Negative values in Figure 5(a) can be explained by the overall energy losses being greater than the energy emitted on those particular days, as the in-vessel systems were not hermetically closed.

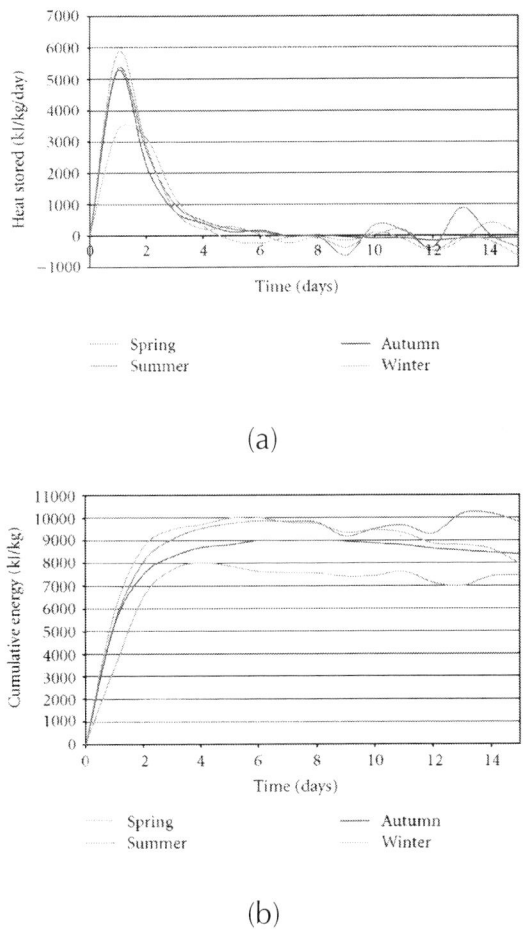

(a)

(b)

Figure 5: (a) Heat stored in compost, and (b) cumulative heat stored by compost; seasonal values.

Cumulative energy values ranged from approximately 7000 to 10 000 kJ kg^{-1}. These values related well to those reported by Ekinci at al. [19] for biosolids and wood chips (8092 kJ kg^{-1}) and Steppa [24] for organic waste (9000–11 000 kJ kg^{-1}), but were lower than that reported by Sobel and Muck [13] for poultry droppings (12 800 kJ kg^{-1}), and by Stainforth [12] for wheat straw (17 600 kJ kg^{-1}). Additionally, they were greater than that for paper mill sludge and poultry manure compost (3649 kJ kg^{-1}), and that for straw and poultry manure compost (5111 kJ kg^{-1}) [23]. Differences in cumulative energy values are due to differences in decomposition rates under different conditions and heat of combustion values of the different composting substrates, which lead any direct comparisons being difficult to arrive upon and ultimately ineffectual. However, it is clear that, as these values are in line with a section of the existing research, they are reliable figures which may be utilised appropriately for subsequent calculations. The main reason for seasonal variation is likely due to the difference in material that is available during that particular season. For example, there will be less nitrogen rich material such as grass cuttings during the autumn and winter months, thus resulting in a lower overall energy content.

Heat Exchanger and Potential Uses for the Gathered Heat

All design parameters are summarised in Table 5, and the pipe layout in the in-vessel unites shown in Figure 6. These layouts provide enough length whilst managing to avoid contact points such as temperature probe holes and exhaust air outtakes. Using this particular pipe dimension leads to the hot and cold fluid exit temperatures that are presented in Table 6. These have been calculated for varying cold water inlet temperatures, due to the potential seasonal variability, and an initial air temperature of 65°C in each case. According to these values, the hot water could be transferred to a storage vessel site office and used to supplement the hot water supply. The predicted daily demand of 180 L day^{-1} could be met in full by the heat exchanger, which runs at a flow rate of 3 L min⁻¹. This would remove the need for gas or electricity to provide hot water. However, in order to meet storage legislation [25] the temperature of the stored water must exceed 60°C at all times. Currently the heated water leaving the tunnel heat exchanger system is at a temperature between 47.3°C and 50°C, depending on the temperature of the cold water entering the system. However, if the water exiting a tunnel is put through another tunnel in which degradation is also underway then higher temperatures can be achieved. For example, calculations based on an initial cold water feed of 0°C and initial hot air temperatures of 65°C result in temperatures of ,0°C ,47.3°C ,60.2°C and 63.7°C by passing water through 0, 1, 2, and 3 tunnels in series. Thus, by passing the same volume of water through 2 or 3 in-vessel tunnels, the required storage temperature of can be achieved. As the hot air is at an approximate temperature of 65°C, then passing the water through more than three in-vessel tunnels has little improved effect and is likely inefficient practice.

Table 5

Inlet temperatures		T_c	0	°C
		T_M	65	°C
		Flow rate	3	L min⁻¹
			0.00005	m³ s⁻¹
Cold fluid flow rate (water pump)		Density water	999	kg m⁻³
		Flow rate	0.04995	kg s⁻¹
		$\dot{m}_c$	0.04995	kg s⁻¹
		$\dot{m}_h$	2000	m³ hr⁻¹
		Density air	1.2	kg m⁻³
Hot fluid flow rate (blower at average capacity)			2400	kg hr⁻¹
		$\dot{m}_h$	0.66667	kg s⁻¹
Specific heat capacities		c_{pc}	4.39	kJ kg⁻¹ K⁻¹
		c_{ph}	1.004	kJ kg⁻¹ K⁻¹
		d_i	0.02465	m
Pipe properties (stainless steel)		d_o	0.02667	m
		Thickness	0.00202	m
		k_w	16.3	W m⁻¹ K⁻¹
Heat transfer coefficients	$h = \dfrac{\Delta Q}{A \times \Delta T \times \Delta t}$	h_i	1200	W m⁻² K⁻¹
		h_o	50	W m⁻² K⁻¹
Heat capacity rates	$C = \dot{m}\, c_p$	C_c	209.2905	W K⁻¹
		C_h	669.3333	W K⁻¹
Heat capacity ratio	$C^* = \dfrac{C_{min}}{C_{max}}$	C_{min}	209.2905	W K⁻¹
		C^*	0.31269	
Fouling capacity (standard)		$R_{w,f}$	0.001	m² K W⁻¹
		$1/h_o$	0.02	
		$d_o/h_i d_i$	0.0009015	
Overall heat transfer coefficient	$\dfrac{1}{U} = \dfrac{1}{h_o} + \dfrac{1}{h_o} + \dfrac{d_o \ln(d_o/d_i)}{2k_w} + \dfrac{d_o}{h_i d_i} + \dfrac{d_o}{A_o d_o}$	Wall	0.0000644	
		Fouling inside	0.0021637	
		Fouling outside	0.002	
		U	39.7936	W m⁻² K⁻¹
Pipe total surface area		A	7.96	M²
NTU	$NTU = \dfrac{UA}{C_{min}}$	NTU	1.5135	
ε	$\varepsilon = \dfrac{1 - \exp[-NTU(1 - C^*)]}{1 - C^* \exp[-NTU(1 - C^*)]}$	eqn top line	0.6466	
		eqn bottom line	0.8895	
		E	0.7270	
		$T_{h,i} - T_{h,o}$	14.7790	
Outlet temperatures	$\varepsilon = \dfrac{C_h(T_{h,i} - T_{h,o})}{C_{min}(T_{h,i} - T_{c,i})} = \dfrac{C_c(T_{c,o} - T_{c,i})}{C_{min}(T_{h,i} - T_{c,i})}$	$T_{h,o}$	50.2250	°C
		$T_{c,o} - T_{c,i}$	47.2520	
		$T_{c,o}$	47.2520	°C
Energy balance check		q	9889.3897	kW
		$T_{h,o}$	50.2250	C
Pipe length needed		Surface area	0.0838	m² m⁻¹
		Pipe length	95.0036	m

Table 6: Outlet fluid temperatures at varying inlet temperatures

Temperature of cold water entering system	0*C	5*c	10*c
Temperature of cold water leaving system	47.3* C	48.6*C	50.0*C
Temperature of hot air leaving system	50.2*C	51.4*C	52.5*C

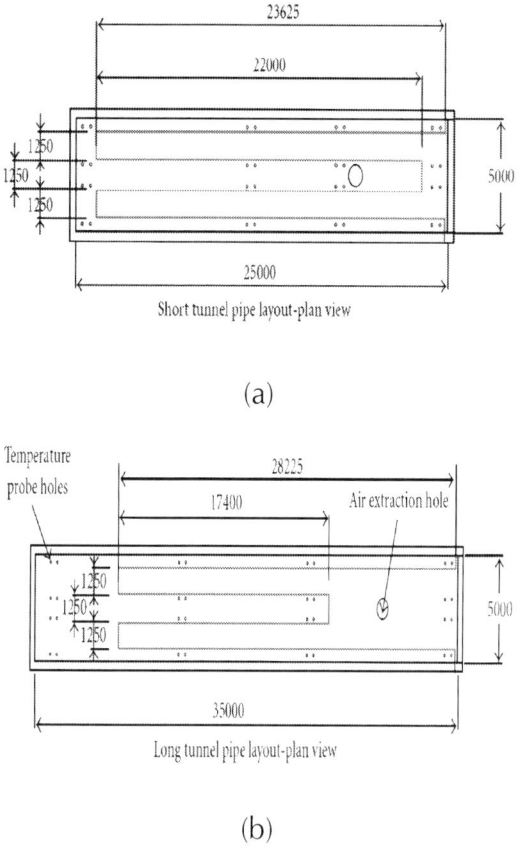

Short tunnel pipe layout-plan view

(a)

Long tunnel pipe layout-plan view

(b)

Figure 6: Suggested pipe layout for short and long in-vessel units.

The main issue with this use is that it requires at least two tunnels to be operating at the same time. If considering the usage over the past year as an accurate gauge then two or more tunnels are running simultaneously on 91% of days and three or more tunnels are running simultaneously on 78% of days (Calculations and graphical solution behind these statements not shown). It can be suggested therefore that domestic hot water supply, which requires a higher temperature for storage, can be supplied at least 78% of the time whilst underfloor spatial heating, as discussed below, can be provided 91% of the time. Another issue is that this system will require additional infrastructure to facilitate the level of control needed to direct the water into the correct tunnels. Two-way switch valves can be installed which could be controlled manually or by computer if required.

If underfloor heating is provided in the new site office then it may be feasible to provide the hot water for the system. Standard operating procedure for underfloor heating is to have an inlet temperature of 55°C and a return flow at 45°C. The water will therefore have to be passed through two in-vessel tunnels in series and so provide 91% of the yearly demand. Standard underfloor systems require a heated water flow rate in the region of 1.55 L min^{-1} [26]. The heat exchanger element has been designed with a flow rate of 3 L min^{-1} so is capable of meeting demand. Water could be stored thus allowing the pumps to be run for fewer hours each day or the pump could be run at a lower capacity thus improving temperature gain further.

Solar Thermal System

The level of potential solar energy supply is 860 kWh m^{-2} year^{-1}, while the predicted hot water demand for the site office of the Deerdykes composting facility is 180 L day^{-1} and the hot water energy demand is 6.8 kWh day^{-1}. These supply and demand parameters led to the quantification of the solar thermal system components as shown in Table 7. The predicted performance of this system is to provide 60% of the offices yearly hot water demand. The remaining 40% will have to be provided by an auxiliary conventional boiler system. Regarding the design for spatial heating, for a heated living area of 200 m^2, the evacuated tube surface area chosen for design was 12.53 m^2, and the store volume 700 L.

Table 7: Solar thermal domestic hot water supply system specification

Parameter	Design value
Solar collector type	Evacuated tube
Solar collector area	4.18 m²
Heat store tank	277 L
Solar circuit piping diameter	8.86 mm
High temperature expansion vessel	18 L

Ground Source Design

The method utilised to size and cost a suitable ground source heat pump system is very simplified and will not provide wholly accurate or reliable answers. This process was carried out in order to provide a comparison of typical cost and performance level of alternate renewable sources and is therefore for guidance purposes only. If a more detailed design is to be carried out a full site survey will be required to determine ground conditions and the levels of thermal insulation provided by the site office.

Cost Benefit Analysis

Table 8 summarises the cost per kWh of energy generated for each of the three possible methods. This cost is based solely on the annual operating cost required to run the system. Capital costs are provided for information and comparative purposes only and are not a factor in the cost per kWh calculation. The fraction of spatial heating provided is based on supplying a standard underfloor heating system with an input temperature of 55°C. Capital costs do not include the underfloor heating system or the heating network itself. These should be similar if not identical for each system.

Table 8: Costs benefit analysis

Method	Compost heat		Solar thermal		Ground source heat pump	
	Domestic hot water supply	Spatial heating	Domestic hot water supply	Spatial heating	Domestic hot water supply	Spatial heating
Capital cost (£)	11 662		4180	13 415	5413	
Operating Cost (£ yr^{-1})	1039		625	940	840	
Fraction Provided	78%	91%	60%	35%	50%	100%
kWh yr^{-1} required	2667	11 760	2667	11 760	2667	11 760
Cost per kWh (£)	0.4994	0.0971	0.3905	0.2284	0.6298	0.0714

From this it is clear that reusing the heat offered by the compost is an economically attractive option when compared to alternate possible renewable sources. It is able to provide the highest proportion of the total hot water requirement of the three systems. Although comparative costs per kWh are greater than for solar thermal regarding domestic hot water supply and greater than ground source heat pump regarding spatial heating, these supplies will have to be boosted to a greater degree by additional conventional boiling.

The capital cost of using the heat from the compost, although large, is also attractive when compared to its competitors. Although the ground source heat pumps capital cost is lower, this initial gain will soon be lost with the high level of additional hot water heating required to provide domestic hot water supply. This cost assumes all 8 tunnels will have a heat exchanger element installed, thus increasing reliability. A cost that has not been included is the valve control system that directs the water into the correct tunnels. This could be controlled manually, but an appropriate computer controlled system will guarantee a reliable supply of water. The expenditure of such a product has not been included.

CONCLUSIONS

The amount of energy that could be obtained from composting at the Deerdykes composting facility near Glasgow has been calculated as between 7000 and 10 000 kJ kg^{-1} for a 15-day composting period. The variations were likely due to seasonal differences in conditions and raw material supply. The results were in line with alternate existing investigations into material of similar composition. This showed that the compost contained a usable amount of energy if it could be gathered.

Methods of extracting the heat were fully researched. A solution of absorbing the heat contained in the expelled air in the tunnel space above the compost was put forward. A bespoke air-water heat exchanger utilising stainless steel piping was designed and specified. The outlet temperature of the water was shown to be 47.3°C. This can be shown to rise to above 60°C if the water is passed through multiple tunnels in series.

Several usable purposes were suggested for this heated water, including contributing to the site offices hot water demand and process optimisation. Adequate levels of heated water were shown to be provided for the domestic hot water and spatial heating supplies for 78% and 91% of the time, respectively.

Installing this system was submitted to preliminary costing in order to calculate a cost per kWh of energy that could be displaced by using the heat of the compost. This cost was then compared with that of specially designed solar thermal and ground source heat pump systems. From this the heat exchanger system could be put into a real world context. The system was found to provide the most reliable supply of the three systems, and to do so at a very competitive price of £0.499 and £0.097 per kWh for domestic hot water supply and spatial heating, respectively. It can therefore be concluded that collecting the waste heat of compost through a heat exchanger is a realistic solution to contributing to energy demand.

Further investigations to maximize the production of heat from in-vessel units are ongoing.

ACKNOWLEDGMENTS

The authors are grateful to Andrew Meldrum and Donald MacBrayne (Deerdykes Composting Facility), for providing data and information required for the completion of this paper. They are also grateful to Paul Steen (Ramboll UK), for providing advice on particular aspects of this paper. They also thank Paolo Pironi (University of Edinburgh) for providing access to thermocouples for further collection of temperature readings outside the composting vessels.

REFERENCES

1. R. T. Haug, Compost Engineering: Principles & Practice, Ann Arbor Science Publishers, Ann Arbor, Mich, USA, 1980.

2. EC, EU Animal By-Products Regulations (2003/31/EEC), European Commission, 2003.

3. B. Antizar-Ladislao, A. J. Beck, K. Spanova, J. Lopez-Real, and N. J. Russell, "The influence of different temperature programmes on the bioremediation of polycyclic aromatic hydrocarbons (PAHs) in a coal-tar contaminated soil by invessel composting," Journal of Hazardous Materials, vol. 144, no. 1-2, pp. 340–347, 2007.

4. L. Allievi, A. Marchesini, C. Salardi, V. Piano, and A. Ferrari, "Plant quality and soil residual fertility six years after a compost treatment," Bioresource Technology, vol. 43, no. 1, pp. 85–89, 1993.

5. M. Chapman and B. Antizar-Ladislao, "Biotic landfill CH4 emission abatement using bio-waste compost as a landfill cover," in Landfill Research Focus, E.C. Lehmann, Ed., Nova Science Publishers, New York, NY, USA, 2007.

6. H. Insam, N. Riddech, and S. Klammer, Microbiology of Composting, Springer, Berlin, Germany, 2002.

7. K. Sipila, A. Johansson, and K. Saviharju, "Can fuel-based ¨ energy production meet the challenge of fighting global warming—a chance for biomass and cogeneration?" Bioresource Technology, vol. 43, no. 1, pp. 7–12, 1993.

8. EEA, "State of renewable energies in Europe 2006," 2008, http: //www.energies-renouvelables.org/observer/stat/baro/ barobilan/barob-ilan6.pdf.

9. B. Antizar-Ladislao and J. L. Turrion-G ´ omez, "Second- ´ generation biofuels and local bioenergy systems," Biofuels, Bioproducts and Biorefining, vol. 2, no. 5, pp. 455–469, 2008.

10. E. Klejment and M. Rosinski, "Testing of thermal properties ´ of compost from municipal waste with a view to using it as a renewable, low temperature heat source," Bioresource Technology, vol. 99, no. 18, pp. 8850–8855, 2008.

11. N. Guljajew and M. Szapiro, "Determining of heat energy volume released by waste during biothermal disposal," in Sbornik Naucznych Robot, pp. 135–141, Akademija Kommunalnowo Chozjajstwa, Moskow, Russia, 1962.

12. A. Stainforth, Cereal Straw, Clarendon, Oxford, UK, 1979.

13. A. T. Sobel and R. E. Muck, "Energy in animal manures," Energy in Agriculture, vol. 2, pp. 161–176, 1983.

14. S. Lekic, Possibilities of Heat Recovery from Waste Composting Process, Centre for Sustainable Development, Department of Engineering, University of Cambridge, Cambridge, UK, 2005.

15. H. Seki and T. Komori, "Packed-column-type heating tower for recovery of heat generated in compost," Journal of Agricultural Meteorology, vol. 48, pp. 237–246, 1992.

16. S.D. Last, D. MacBrayne, and A. J. MacArthur, "Deedykes Composting Facility: a case study of the conversion of a conventional activated sludge sewage works to in-vessel composting, with slude co-composting facility," in Proceedings of Kalmar Eco-Tech and The 2nd Baltic Symposium on Environmental Chemistry, Kalmar, Sweden, 2005.

17. DGS, Planning and Installing Solar Thermal Systems: A Guide for Installers, Architects and Engineers, James & James, London, UK, 2005.

18. P. S. Doherty, S. Al-Huthaili, S. B. Riffat, and N. Abodahab, "Ground source heat pump—description and preliminary results of the Eco House system," Applied Thermal Engineering, vol. 24, no. 17-18, pp. 2627–2641, 2004.

19. K. Ekinci, H. M. Keener, and D. Akbolat, "Effects of feedstock, airflow rate, and recirculation ratio on performance of composting systems with air recirculation," Bioresource Technology, vol. 97, no. 7, pp. 922–932, 2006.

20. R. K. Shah and D. P. Sekulic, Fundamentals of Heat Exchanger Design, John Wiley & Sons, Hoboken, NJ, USA, 2003.

21. R. D. Heap, Heat Pumps, E. & F.N. Spon, London, UK, 1979.

22. K. Ekinci, Theoretical and Experimental Study on the Effects of Aeration Strategies on the Composting Process, Department of Agricultural Engineering, The Ohio State University, Columbus, Ohio, USA, 2001.

23. E. Harper, F. C. Miller, and B. J. Macauley, "Physical management and interpretation of an environmentally controlled composting ecosystem," Australian Journal of Experimental Agriculture, vol. 32, pp. 657–667, 1992.

24. M. Steppa, " Two options for energy recovery from waste biomass," Maszyny i Ciagniki Rolnicze, no. 3, 1988.

25. HSE, "Legionnaires disease: essential information for providers of residential accommodation," Health and Safety Executive, 2003.

26. Oxyvent, "The Oxyvent Tank: Underfloor heating / Radiant heating," 2009, http://www.oxyvent.com/underfloorheating.php.

Chapter 7

Analytical Modelling of a Spray Column Three-phase Direct Contact Heat Exchanger

Hameed B. Mahood[1,2], Adel O. Sharif[1], Seyed Ali Hosseini[1], and Rex B. Thorpe[1]

[1]Centre for Osmosis Research and Applications (CORA), Chemical and Process Engineering Department, Faculty of Engineering and Physical Sciences, University of Surrey, Guildford GU2 7XH, UK

[2]University of Misan, Misan, Iraq

ABSTRACT

An analytical model for the temperature distribution of a spray column, three-phase direct contact heat exchanger is developed. So far there were only numerical models available for this process;

however to understand the dynamic behaviour of these systems, characteristic models are required. In this work, using cell model configuration and irrotational potential flow approximation characteristic models has been developed for the relative velocity and the drag coefficient of the evaporation swarm of drops in an immiscible liquid, using a convective heat transfer coefficient of those drops included the drop interaction effect, which derived by authors already. Moreover, one-dimensional energy equation was formulated involving the direct contact heat transfer coefficient, the holdup ratio, the drop radius, the relative velocity, and the physical phases properties. In addition, time-dependent drops sizes were taken into account as a function of vaporization ratio inside the drops, while a constant holdup ratio along the column was assumed. Furthermore, the model correlated well against experimental data.

INTRODUCTION

A direct contact heat exchanger is a highly effective device for transferring heat between two immiscible fluids while they are flowing co-currently or countercurrently inside a column. The main feature of the direct contact heat exchanger is that it permits a confident contact between a hot fluid and a cold fluid. However, a number of different methods have been used to define the type of direct contact heat exchanger, including layer type, where the hot fluid is stagnant while the cold fluid flows on top, and a spray type, where one of two fluids is injected into the other. Generally, there are two types of spray column, depending on which injection technique is being used: an integrated type and a split type. In the former, the cold fluid is dispersed from the bottom of the column into a hot fluid, whereas in the second one, the hot fluid is pumped in countercurrently with the flowing cold fluid [1].

A direct contact heat exchanger has several advantages over surface heat exchangers [2], such as eliminating metallic heat transfer surface between fluids which are prone to corrosion and fouling, as well as increasing the heat transfer resistance. It can be operated at very low temperature differences or heat transfer driving forces and allows lower mass flow rates of transferring fluids, convenient separation of the fluids, and a high heat transfer coefficient (about 20–100 times than single phase or surface type heat exchanger) [3]. Therefore, it can be found in several industrial applications, such as water desalination by freezing, geothermal power energy production, crystallization, waste heat recovery, energy storage systems, solar power energy, and emergency cooling of chemical and nuclear reactors [4]. Considerable attention has been paid to the area of direct contact heat exchangers in recent years, particularly when change of phase takes place, but most of the efforts have been focused on the evaporation of single drops or condensation of single bubbles.

A very limited number of experimental, theoretical, and numerical investigations have dealt with the evaporation of multidrops in an immiscible liquid media. In addition, the available works in this area have concentrated mainly on single parameters affecting the phenomena of heat transfer in direct contact heat exchangers, such as a heat transfer coefficient and holdup ratio. Sideman and Gat [5] studied experimentally the operating characteristics of a spray column using a pentane-water system and calculated the volumetric heat transfer coefficient and average holdup ratio as a function of pentane and water superficial velocities. The temperature profiles of the dispersed (kerosene) and continuous phases (water) in a spray-column direct contact heat exchanger, were investigated both experimentally and theoretically by Letan and Kehat [6]. They proposed that the mechanism of heat transfer inside the spray column could be characterized by five regions: wake growth, intermediate, continuous wake shedding, mixing, and coalesce regimes, respectively. In addition, they observed that the heat exchange between the two phases in both heating and cooling processes occurs only in the certain zones wake growth, continuous wake shedding, and mixing region. The characteristics of the liquid-liquid spray-column direct contact heat exchanger were studied by Plass et al. [7] experimentally. The volumetric heat transfer coefficient was found and they correlated their experimental results of volumetric heat transfer as a function of holdup ratio. By using a numerical technique Coban and Boehm [8], Jacobs and Golafshani [9] and Brickman and Boehm [10] developed a one-dimensional numerical model to study the flow and the heat transfer of a spray column direct contact heat exchanger. The mass, momentum, and energy equations of two-component two-phase flow were solved and the temperature profile, total heat transfer, and holdup ratio were calculated. Several heat transfer models for interfacial heat transfer between the dispersed and continuous phases were examined by [9] and they found that the conduction controlled heat transfer is dominated by small diameter liquid drops. While, the possibility of increasing a spray column direct contact heat transfer output was included in the analysis of Brickman and Boehm [10]. They found that decreasing the dispersed phase inlet temperature while maintain a continuous phase inlet temperature produced an increase of 10–20% in heat transfer inside the column. Also, the optimum conditions can be achieved when using the dispersed phase at or very near to its saturation temperature. The effect of the superficial velocity and the initial temperature of the continuous phase and the dispersed phase on the volumetric heat transfer coefficient of an n-pentane-water, three-phase, direct-contact heat exchanger were investigated by Peng et al. [3] experimentally and theoretically. Their results indicated that the volumetric heat transfer coefficient increases with increasing initial temperature and superficial velocity of continuous phase, while the superficial dispersed phase velocity had no effect. More accurate numerical model is carried out by Kang et al. [11] to study the heat transfer characteristics of a spray column direct contact heat exchanger. A two-dimensional axisymmetric two-component flow model was developed and they found the injection velocity of dispersed phase has a dominated effect more than other parameters and the volumetric heat transfer coefficient is uniform until the middle column is reached.

So far there were only numerical models available which deal with the direct contact heat transfer process in a spray column direct contact heat exchanger; however to understand the dynamic behaviour of these systems, characteristic models are required. In this investigation, an analytical model is developed to study the characteristic heat transfer of a spray-column direct contact heat exchanger. The analysis is based on solving the one-dimensional energy equation for two-component two-phase flow using a cell model. The direct contact heat transfer coefficient, drag coefficient, and relative velocity of multidrops' evaporation during their flow in the countercurrent with hot water were determined and the temperature distribution along the column height was found. It is important here to note that no analytical model is currently available to describe the temperature distribution of the temperature inside the direct contact column.

THEORETICAL MODELLING

Initially, due to the large diameter of the column in comparison to the diameter of the drops, and because there is no circulation zone inside the column, the flow inside the column can be assumed to be a one-dimensional flow [9]. On the other hand, immiscibility between the countercurrent flow phases means the mass flow rate of both phases stays constant along the column because neither the dispersed phase or the continuous phasen goes into solution in the other phase.

The continuity equations for countercurrent flow of the phases can be written as [8]

$$\dot{m}_d = \rho_d A\alpha U_d,$$

(1)

$$\dot{m}_c = \rho_c A(1-\alpha)U_c.$$

(2)

For one-dimensional, steady-state flow, the energy equations for both phases can be written as

$$\frac{d}{dz}\left[\rho_d\alpha\, U_d H_d\right] = -\frac{Q}{V},$$

(3)

$$\frac{d}{dz}\left[\rho_c\left(1-\alpha\right)U_c\,H_c\right]=\frac{\eta Q}{V}.$$

(4)

Substituting (1) and (2) into (3) and (4), respectively, yields

$$\frac{dH_d}{dz}=-\frac{A}{\dot{m}_d}\frac{Q}{V},$$

(5)

$$\frac{dH_c}{dz}=\frac{A}{\dot{m}_c}\frac{\eta Q}{V},$$

(6)

Where Q represents the heat transfer from the continuous phase to the dispersed phase and η denotes the ratio of the heat transfer from the continuous phase divided by the heat transfer to the dispersed phase, when no heat is lost to the surroundings:

$$\eta=1.$$

(7)

Equations (5) and (6) can be written in terms of the phase temperature, as:

$$\frac{dT_c}{dz}=-\frac{Q}{\dot{m}_c c_{pc}}\frac{A}{V},$$

(8)

$$\frac{dT_d}{dz}=\frac{Q}{\dot{m}_d c_{pd}}\frac{A}{V},$$

(9)

Where

$$H_c=c_{pc}T_c,$$

$$H_d=c_{pd}T_d,$$

(10)

Where H_d and H_c denote the enthalpies of the dispersed phase and the continuous phase, respectively, and (A / V) is the total heat transfer area per unit volume. Plass et al. [7] suggested the following expression for this parameter and for spherical drops:

$$\frac{A}{V} = \frac{6\alpha}{d},$$

(11)

Where α denoted the holdup ratio or the ratio between the volumes of dispersed phase in the column and the total volume (dispersed and continuous phase volume) and d is the column diameter.

Substituting (11) into (8) and (9) results in

$$\frac{dT_c}{dz} = -\left(\frac{6\alpha}{d}\right)\frac{Q}{\dot{m}_c c_{pc}},$$

(12)

$$\frac{dT_d}{dz} = \left(\frac{6\alpha}{d}\right)\frac{Q}{\dot{m}_d c_{pc}}.$$

(13)

The heat transfer from the continuous phase to the dispersed phase Q can be found as

$$Q = hA\Delta T,$$

(14)

where h is the direct contact heat transfer coefficient. Brikman and Boehm [10] and others used the heat transfer coefficient for a single evaporation drop and multiplied it by the number of drops to include the void fraction in their analysis. However, it is more useful to use an expression for heat transfer coefficient related to the multidrops or holdup ratio inside the column. Actually, the phenomenon of heat transfer to the droplets, including the droplet phase change, is quite complex due to coalescing or breaking down of the droplets along the column height. In addition, droplet shapes might change due to evaporation from spherical to ellipsoidal and finally to spherical-cup shapes. Therefore, an assumption should be made to allow simplification of the complexity of such phenomena. To this end, it is useful to assume that the droplets remain spherical along the direct contact heat transfer process, and no coalescing nor breaking down occurs for the evaporation droplets and finally that there is a constant droplet number. Assume the droplets swarm to be spherical in shape and that they move in potential flow fields with a constant drop radius, and by solving the steady-state energy equation for the spherical coordinate using a potential flow configuration for the velocity components, and a cell model assumption, Mahood (2012) [12] has found the heat transfer coefficient in terms of Nu number, for the multidrops evaporation in an immiscible liquid as

$$\text{Nu} = \frac{4}{\sqrt{6\pi}} \cdot \left(\frac{\alpha + 0.5}{1 - \alpha} \right)^{0.5} \cdot \text{Pe}^{0.5}.$$

(15)

The convective heat transfer coefficient in (15) was derived by solving a steady-state energy equation in spherical coordinates, which involved velocity components. These components of the velocity were found from the solution for the potential flow field around the two-phase bubbles. Because no viscous or actual solution dealing with the evaporation of the two-phase bubble is available to compare with the potential flow solution, Isenberg and Sideman [13], Moalem et al. [14], and Moalem et al. [15] used the flowing velocity factor, in which the solution is based on the assumption that the potential flow is converted to an actual or viscous solution:

$$k_v = 0.25 \ \text{Pr}^{-1/3},$$

(16)

and for pure potential flow

$$k_v = 1.$$

(17)

Therefore, (15) becomes:

$$\text{Nu} = \frac{4}{\sqrt{6\pi}} \cdot \left(\frac{\alpha + 0.5}{1 - \alpha} k_v \right)^{0.5} \text{Pe}^{0.5}.$$

(18)

The heat transfer coefficient is related to the slip or relative velocity of the two phases in the column, so it is important to derive the relative velocity expression for the countercurrent two-phase flow inside the column. Using a cell model as shown in Figure 1, the potential velocity of the drops in the swarm was given by Milne-Thomson [16] and used by Cai and Wallis [17] as follows:

$$\phi = \frac{\cos\theta}{b^3 - a^3}\left[\left(a^3 U - b^3 v\right)r + \frac{a^3 b^3}{2r^2}\left(U - v\right)\right],$$

$$(19)$$

where U and v are represented velocity of inner and outer cell boundaries, respectively.

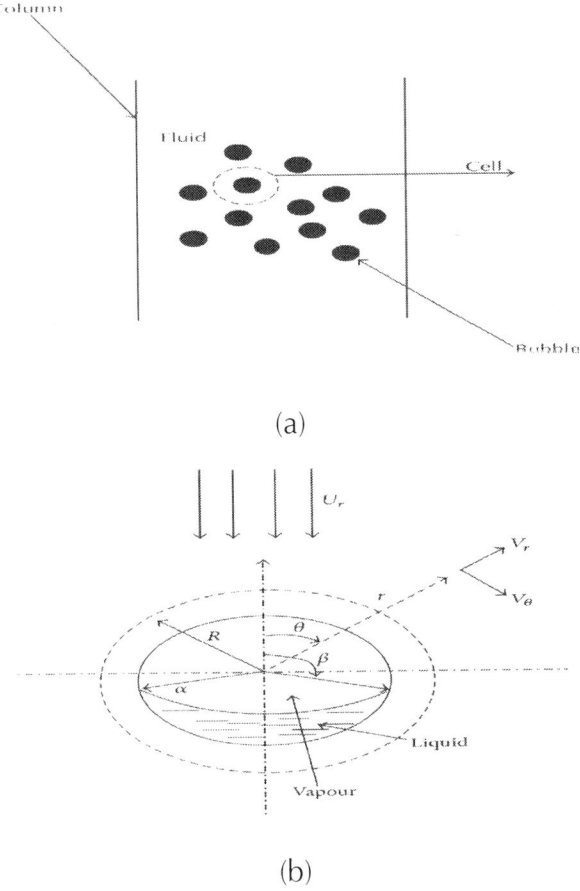

(a)

(b)

Figure 1: Schematic representation of the cell model.

For v=0, (19) reduces to the expression given by Lamb [18] and used by Kendoush [19] as

$$\phi = \frac{U}{b^3 - a^3}\left[a^3 r + \frac{a^3 b^3}{2r^2}\right]\cos\theta.$$

(20)

And for $\alpha = (a/b)^3$, (20) becomes:

$$\phi = \frac{U}{(1-\alpha)}\left[\alpha r + \frac{1}{2}\left(\frac{a^3}{r^2}\right)\right]\cos\theta.$$

(21)

The velocity components of potential flow can be found using (21) as

$$V_r = -\frac{\partial\phi}{\partial r} = \frac{U}{(1-\alpha)}\left[\alpha - \left(\frac{a}{r}\right)^3\right]\cos\theta,$$

$$V_\theta = -\frac{\partial\phi}{r\partial\theta} = \frac{U}{(1-\alpha)}\left[\alpha + \frac{1}{2}\left(\frac{a}{r}\right)^3\right]\sin\theta,$$

(22)

Where V_r and V_θ are the radial and tangential velocity components.

Equations in (22) are different from these equations used by Marrucci [20] and Kendoush [21, 22] when they analysed the problem of gas bubble swarm, which might lead to an error in their results.

Using a method similar to that suggested by Kendoush [21, 22], the drag force acting of the drops swarm evaporating in an immiscible liquid has been found by Mahood [12] as

$$F_D = -8\pi\mu a U f(\alpha)\, f(\mu)\, f_{Re},$$

(23)

Where

$$f(\alpha) = \frac{(\alpha + 0.5)}{(1 - \alpha)^2},$$

$$f(\mu) = \frac{3\mu_i + 2\mu_o}{2(\mu_i + \mu_o)},$$

$$f_{Re} = 1 + 0.15Re^{0.687},$$

(24)

and the drag coefficient is

$$C_D = \frac{48}{Re} \frac{(\alpha + 0.5)}{(1 - \alpha)^2} \left(1 + 0.15Re^{0.687}\right).$$

(25)

The power of both drag force and buoyancy force should be equal to the rate of the total kinetic energy [23]

$$(F_B + F_D)U = \frac{d}{dt}(K \cdot E)$$

$$= (M + \overline{m}C_v)U\frac{dU}{dt},$$

(26)

where the term $(\overline{m}C_v)$ denotes the added mass, which is derived for a swarm of spherical vapour bubbles evaporation in an immiscible liquid medium by Mahood [12] and Mahood et al. [24] as

$$\overline{M} = \overline{m}C_v = \frac{4}{3}\pi\rho_c a^3 \frac{f_1(\alpha)}{(1 - \alpha)^2},$$

(27)

Where

$$f_1(\alpha) = \frac{1}{2} + \frac{1}{2}\alpha - \alpha^2.$$

(28)

And F_B is the buoyancy force, which can be written as:

$$F_B = \frac{4}{3}\pi a^3 (1 - \alpha) \rho_c g$$

(29)

Substituting (23) and (29) into (26), and ignoring bubbles mass, in comparison with added mass that is, $M \ll \bar{m}C_v$, (26) becomes

$$g\frac{(1 - \alpha)}{f_2(\alpha)} - \frac{6}{\rho_c a^2}\mu U \frac{f(t)}{f_2(\alpha)} = \frac{dU}{dt},$$

(30)

Where

$$f_2(\alpha) = \frac{f_1(\alpha)}{(1 - \alpha)^2} = \frac{0.5 + 0.5\alpha - \alpha^2}{(1 - \alpha)^2},$$

(31)

$$f(t) = f(\alpha) f(\mu) f_{Re}$$

(32)

Equation (30) can be solved in a similar way to Joseph's [23] and Concha's [25], which results in

$$U = \frac{g(1 - \alpha)\rho_c a^2}{6\mu f(t)}$$

$$\times \left[1 - \exp\left(-\frac{6\mu}{\rho_c a^2} \cdot \frac{f(t)}{f_2(\alpha)} \cdot t\right)\right].$$

(33)

For steady-state velocity or terminal velocity, with assuming a drop with a rigid wall due to the contiments, which leads to $f(\mu) = 3/2$, (33), (33) becomes

$$U = \frac{g \rho_c a^2 (1 - \alpha)^3}{9\mu f_{Re} (\alpha + 0.5)}.$$

(34)

For $\alpha \to 0$, (34) reduces to an equation for a single drop as

$$U\left(o\right) = \frac{2g\rho_c a^2}{9\mu\, f_{\mathrm{Re}}}.$$

(35)

The ratio of bubbles warm velocity to the single bubble velocity can be found as:

$$\frac{U\left(\alpha\right)}{U\left(o\right)} = \frac{1}{2}\left[\frac{(1-\alpha)^3}{(\alpha+0.5)}\right],$$

(36)

Or

$$U_r = \frac{U\left(o\right)}{2}\left[\frac{(1-\alpha)^3}{(\alpha+0.5)}\right].$$

(37)

Equation (15), now, can be written in terms of the heat transfer coefficient and the relative phases velocity as

$$h = \frac{2\sqrt{2}\,k_c\beta^{0.5}}{\sqrt{6\pi}\ \sqrt{a}}\left(\frac{U_r}{\epsilon}\,k_v\right)^{0.5},$$

(38)

Where

$$\mathrm{Nu} = \frac{2ah}{k_c},$$

$$\mathrm{Pe} = \left(\frac{2a\,U_r}{\epsilon}\right).$$

(39)

Using (37), (38) converts to

$$h = \frac{2k_c}{\sqrt{6\pi}}\left(\frac{\beta U\left(o\right)}{a\epsilon}\,k_v\right)^{0.5}\left(\frac{(1-\alpha)^{3/2}}{(\alpha+0.5)^{1/2}}\right).$$

(40)

The evaporating drop radius is time dependent and can be found using the simple relation given by Wanchoo and Rina [26] as follows:

$$a = a_o(1 + x(s - 1))^{1/3},$$

(41)

where x and s denote the vaporization ratio and the dispersed phase liquid density, to its vapour density respectively.

Substituting (41) into (40) and substituting the result into (14), then substituting the result of this into (12) and (13) after completing the integration we obtain the following

$$T_c = T_{co}$$

$$-\left[\left(\frac{12}{\sqrt{6\pi}}\right)\left(\frac{\alpha k_c A \, \Delta T}{d \, \dot{m}_c c_{pc}}\right)\left(\frac{\beta U_o}{a\epsilon}\right)^{0.5}\left(\frac{(1-\alpha)^{3/2}}{(\alpha+0.5)^{1/2}}\right)\cdot z\right],$$

(42)

$$T_d = T_{do}$$

$$+\left[\left(\frac{12}{\sqrt{6\pi}}\right)\left(\frac{\alpha k_c A \Delta T}{d \, \dot{m}_d c_{pd}}\right)\left(\frac{\beta U_o}{a\epsilon}\right)^{0.5}\left(\frac{(1-\alpha)^{3/2}}{(\alpha+0.5)^{1/2}}\right)\cdot z\right].$$

(43)

RESULTS AND DISCUSSION

The analytical results for temperatures distribution of both dispersed and continuous phases along the spray column in the direct contact heat exchanger are compared with the experimental data given by Olander et al. [27] to verify the theoretical model results. According to (42) and (43), only the initial phases temperatures and mass flow rates are needed to obtain the temperature distribution along the spray column, assuming a value for holdup ratio within the range calculated by Golafshani [28]. He mentioned that the holdup ratio increases very slowly inside the column due to increasing drop diameters, which is assumed to be a constant through his numerical analysis. In fact, the range of the holdup ratio increase was from

0.14 to 0.165 only. In current analysis, the drop diameters taken as a variable depend on the vaporization rate or the ratio inside the drop. Accordingly, and because of unavailable expression being available to describe this factor during the evaporation of drops or condensation of bubbles in an immiscible liquid, the experimental data was fitted to that given by Sideman and Hirsch [29], which was for a two-phase bubble condensation in an immiscible liquid media. Sideman and Hirsch [29] and others found much similarity in the heat transfer mechanism between the cases of evaporation drops and condensation bubbles in immiscible liquid media.

Figures 2 and 3 show the variation of dispersed and continuous phase temperatures and the spray column height at different initial phases temperatures and flow rates, with an initial drop radius equal to 2 mm and 1.6 mm. It can be clearly seen, that three visible regions appear on these figures, which agrees with [1]. The first region is at the phases' entrance, where the temperature difference is at its maximum between the phases. In this region a high increase in dispersed temperature occurs, while nearly a constant temperature in the continuous phase. This zone covers a very short length of the column (about 1 m), and it seems independent of the operational column parameters.

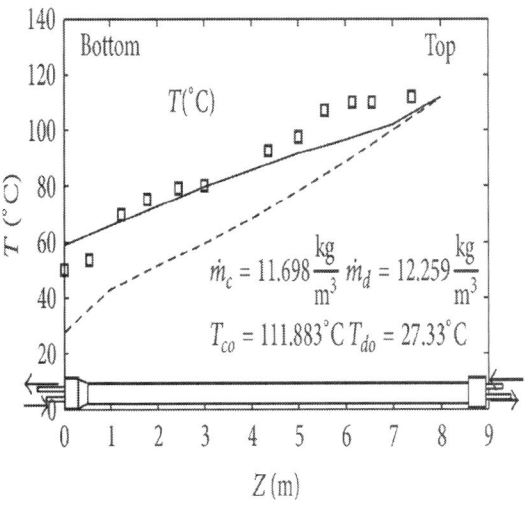

(a)

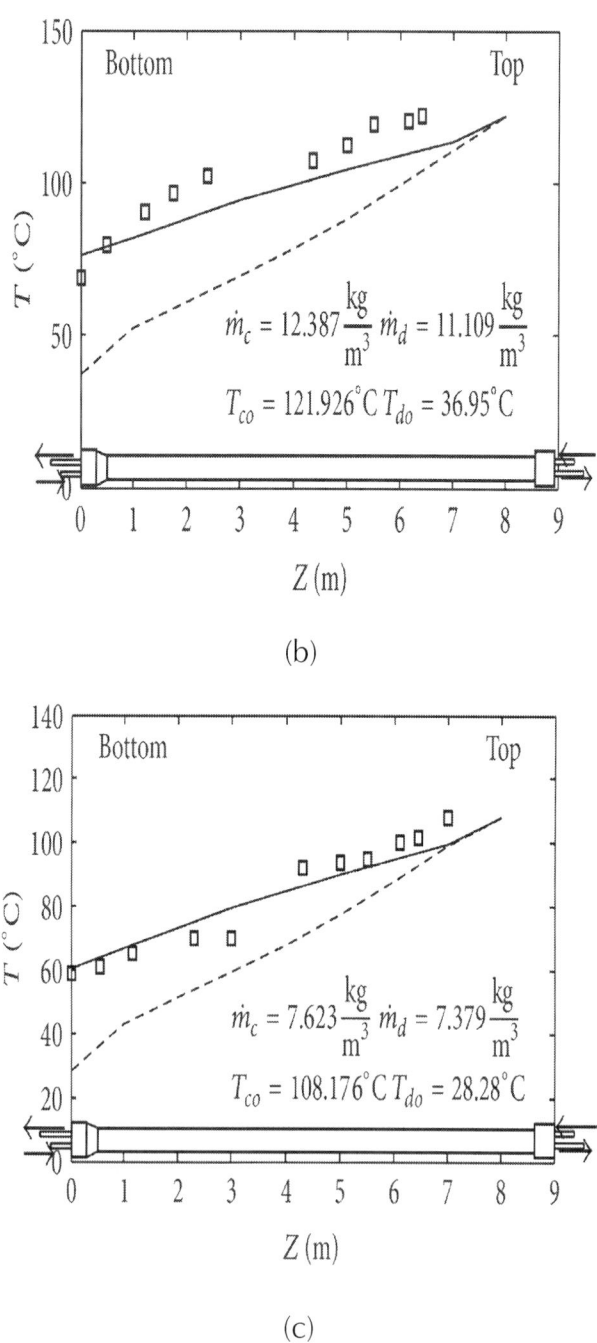

(b)

(c)

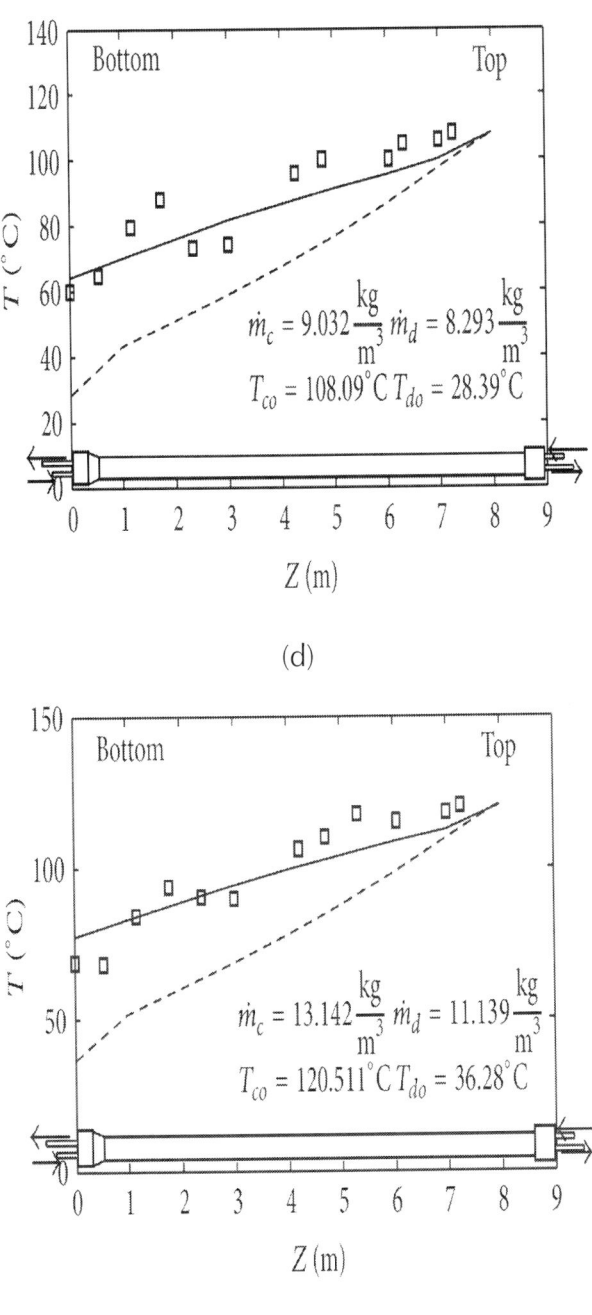

(d)

(e)

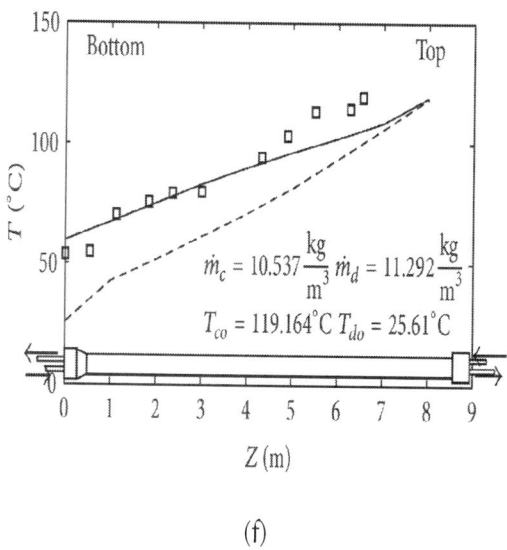

(f)

Figure 2: The variation in continuous phase (—) and dispersed phases (- - - -) temperatures with direct contact heat exchanger height for a=2mm.

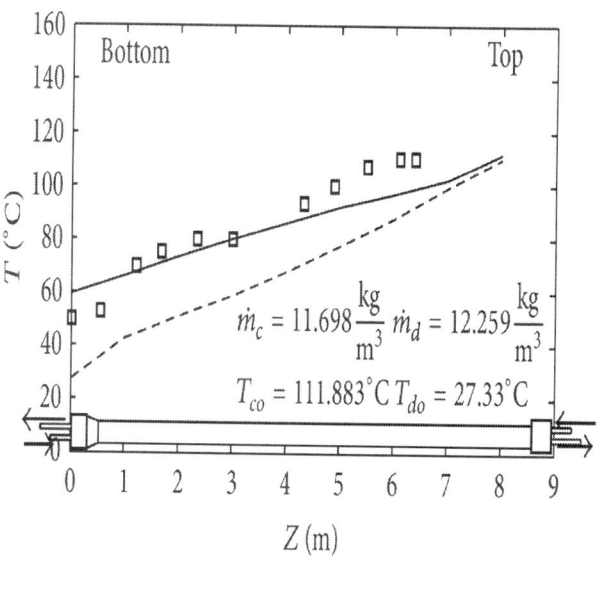

(a)

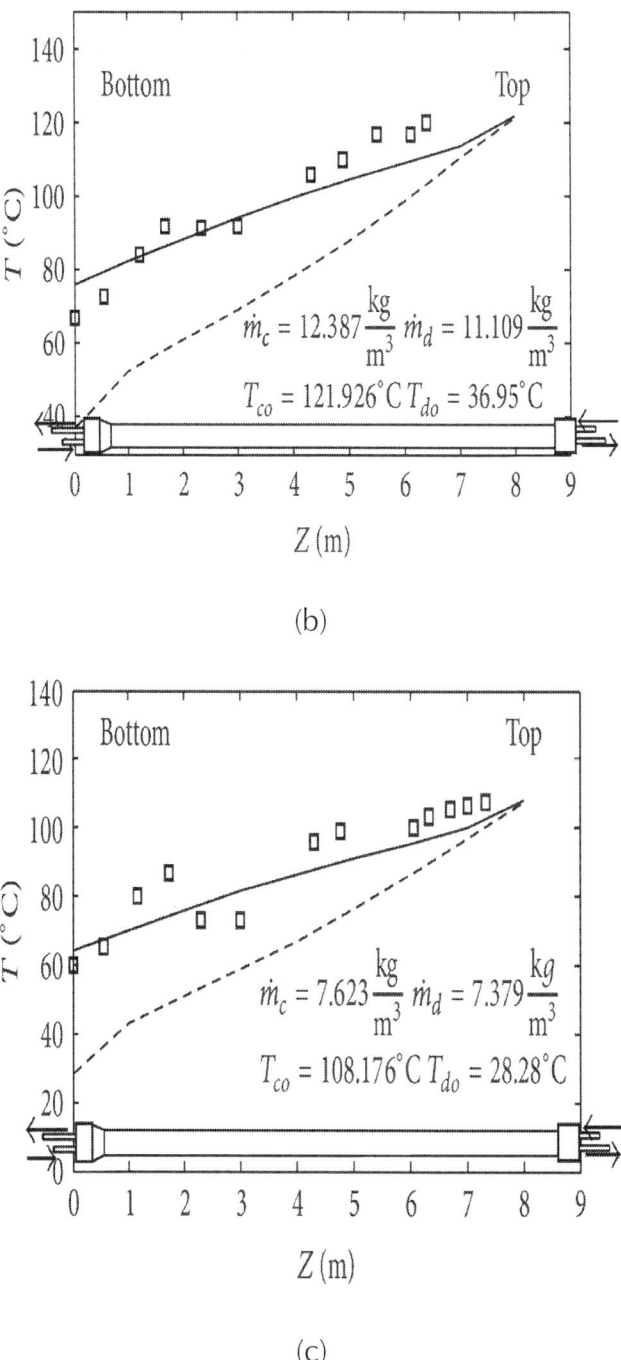

(b)

(c)

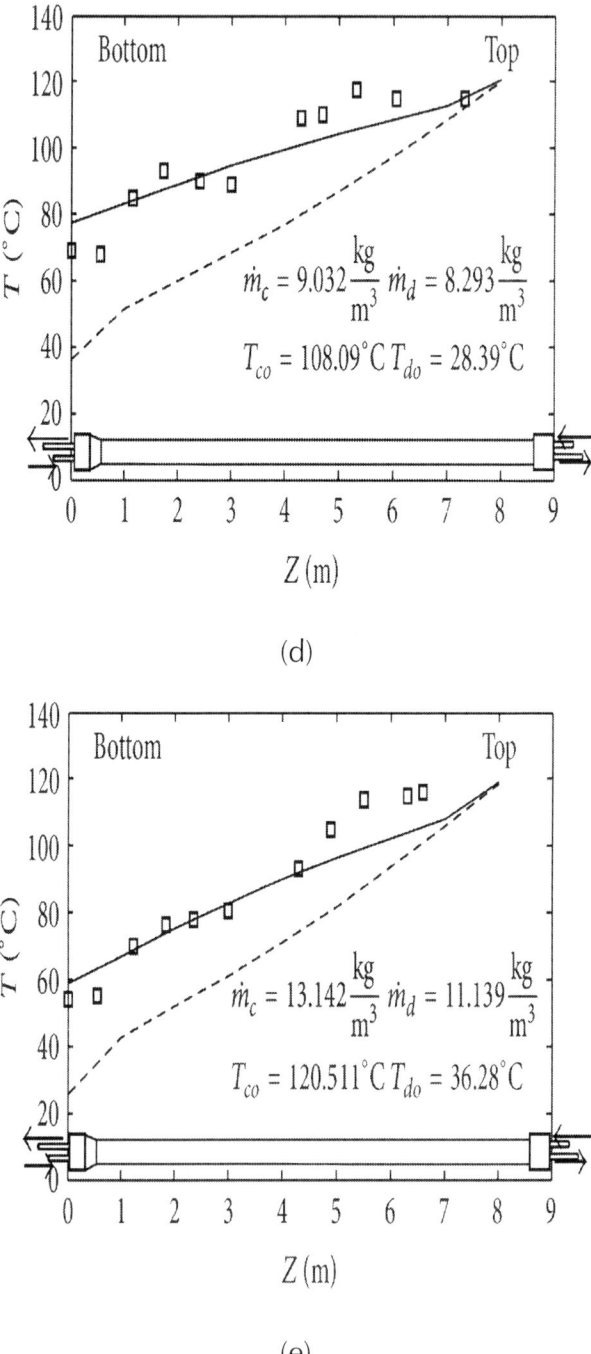

$$\dot{m}_c = 9.032\,\frac{kg}{m^3}\quad \dot{m}_d = 8.293\,\frac{kg}{m^3}$$

$$T_{co} = 108.09\,^\circ C\quad T_{do} = 28.39\,^\circ C$$

(d)

$$\dot{m}_c = 13.142\,\frac{kg}{m^3}\quad \dot{m}_d = 11.139\,\frac{kg}{m^3}$$

$$T_{co} = 120.511\,^\circ C\quad T_{do} = 36.28\,^\circ C$$

(e)

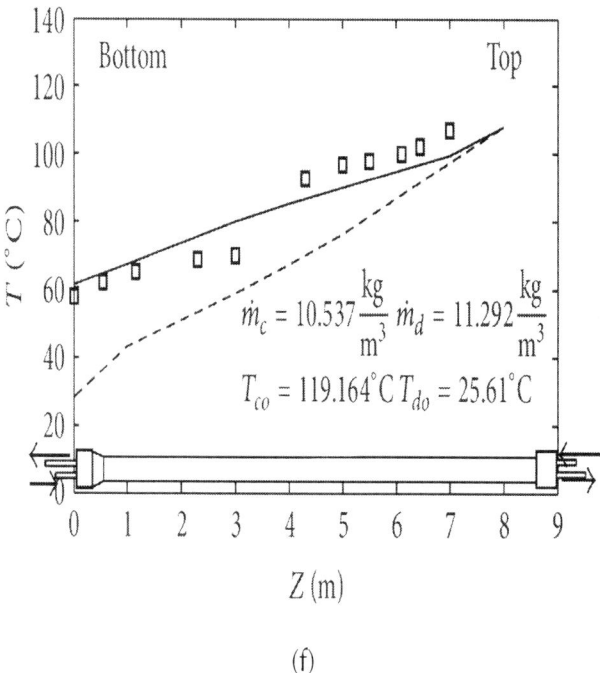

(f)

Figure 3: The variation in continuous phase (—) and dispersed phases (- - - -) temperatures with direct contact heat exchanger height for a=1.6mm.

The second zone starts after nearly one metre of the column height and has a slow heat exchange between the phases. According to the results, it seems to cover a wide range of the column height. The final zone, which represents the last opportunity for heat exchange, and at this region the temperature difference decreases to minimum. Therefore, it seems to be an extension of the second region effect.

From the results, good agreement seem to have been achieved between the analytical model results and the experimental results of Olander et al. [27] for all figures. The main reason or the divergence in the results of the dispersed phase from the experimental data is that the devices basically measuring the water (continuous phase) temperature instead of the dispersed (two-phase bubbles). Therefore, sometime an averaging technique is used to obtain the mixed temperature [8].

In addition, the maximum divergence between the present analytical model and the experimental data occurs at the bottom of the column especially when the dispersed phase is injected bellow the saturation temperature. In this case the evaporation delayed as drops would need to reach their saturation temperature before the evaporation starts, meaning that heat will be consumed within the drops. The maximum error in this region is nearly 12.18%.

CONCLUSIONS

An analytical model has been developed for the temperature distribution of a spray-column, three-phase direct contact heat exchanger. According to the model developed in this work, we have shown that it is reasonable to assume a constant holdup ratio along the direct contact column, which is in accordance with other numerical models in the literature such as Cabon and Boehm [8], Jacobs and Golafshani [9], and Kang et al. [11]. Even more, it has been shown that the vaporization ratio of drops is an influential parameter in the heat exchange process. As expected, we have proved that the rate of heat transfer increases with decreasing drops size. Since our model correlated well with experimental results in the literature (e.g., Olander et al. [28]) we expect that this model will enhance optimisation of practical applications in areas such as the production of electrical power from geothermal hot brine, extracting energy from salt gradient solar pond, and the process of heating water by heat collecting working fluids.

REFERENCES

1. L. Tadrist, P. Seguin, R. Santini, J. Pantaloni, and A. Bricard, "Experimental and numerical study of direct contact heat exchangers," International Journal of Heat and Mass Transfer, vol. 28, no. 6, pp. 1215–1227, 1985.

2.   M. Song, A. Steiff, and P. M. Weinspach, "Direct-contact heat transfer with change of phase: a population balance model," Chemical Engineering Science, vol. 54, no. 17, pp. 3861–3871, 1999.

3.   Z. Peng, W. Yiping, G. Cuili, and W. Kun, "Heat transfer in gas-liquid-liquid three-phase direct-contact exchanger," Chemical Engineering Journal, vol. 84, no. 3, pp. 381–388, 2001.

4.   F. Dammel and H. Beer, "Heat transfer from a continuous liquid to an evaporating drop: a numerical analysis," International Journal of Thermal Sciences, vol. 42, no. 7, pp. 677–686, 2003.

5.   S. Sideman and Y. Gat, "Direct contact heat transfer with change of phase: spray column studies of a three—phase heat exchanger," AIChE Journal, vol. 12, no. 2, pp. 296–303, 1966.

6.   R. Letan and E. Kehat, "The mechanism of heat transfer in spray column heat exchanger," AIChE Journal, vol. 14, no. 3, pp. 398–405, 1968.

7.   S. B. Plass, H. R. Jacobs, and R. F. Boehm, "Operational characteristics of spray column type direct contact preheater," AIChE Symposium Series—Heat Transfer, vol. 75, no. 189, pp. 227–234, 1979.

8.   T. Coban and R. Boehm, "Performance of a three-phase, spray-column, direct-contact heat exchanger,"Journal of Heat Transfer, vol. 111, no. 1, pp. 166–172, 1989.

9.   H. R. Jacobs and M. Golafshani, "Heuristic evaluation of the governing mode of heat transfer in a liquid-liquid spray column," Journal of Heat Transfer, vol. 111, no. 3, pp. 773–779, 1989.

10.   R. A. Brickman and R. F. Boehm, "Maximizing three-phase direct-contact heat exchanger output,"Numerical Heat Transfer A, vol. 26, no. 3, pp. 287–299, 1994.

11.   Y. H. Kang, N. J. Kim, B. K. Hur, and C. B. Kim, "A numerical study on heat transfer characteristics in a spray column direct

contact heat exchanger," KSME International Journal, vol. 16, no. 3, pp. 344–353, 2002.

12. H. B. Mahood, "Theoretical modelling of three-phase direct contact spray column heat exchanger," AMPhil-PhD Transfer Report, University of Surrey, 2012.

13. J. Isenberg and S. Sideman, "Direct contact heat transfer with change of phase: bubble condensation in immiscible liquids," International Journal of Heat and Mass Transfer, vol. 13, no. 6, pp. 997–1011, 1970.

14. D. Moalem, S. Sideman, A. Orell, and G. Hetsroni, "Direct contact heat transfer with change of phase: condensation of a bubble train," International Journal of Heat and Mass Transfer, vol. 16, no. 12, pp. 2305–2319, 1973.

15. D. Moalem-Maron, M. Sokolov, and S. Sideman, "A closed periodic condensation-evaporation cycle of an immiscible, gravity driven bubble," International Journal of Heat and Mass Transfer, vol. 23, no. 11, pp. 1417–1424, 1980.

16. L. M. Milne-Thomson, Theoretical Hydrodynamics, Macmilan, London, UK, 5th edition, 1972.

17. X. Cai and G. B. Wallis, "A more general cell model for added mass in two-phase flow," Chemical Engineering Science, vol. 49, no. 10, pp. 1631–1638, 1994.

18. H. Lamb, Hydrodynamics, Cambridge University Press, 6th edition, 1932.

19. A. A. Kendoush, "Theory of convective drop evaporation in direct contact with an immiscible liquid,"Desalination, vol. 169, no. 1, pp. 33–41, 2004.

20. G. Marrucci, "Rising velocity of a swarm of spherical bubbles," Industrial and Engineering Chemistry Fundamentals, vol. 4, no. 2, pp. 224–225, 1965.

21. A. A. Kendoush, "Hydrodynamic model for bubbles in a swarm," Chemical Engineering Science, vol. 56, no. 1, pp. 235–238, 2001.

22. M. Worner, "A compact introduction to the numerical modelling of multiphase flow,"Forschungszentrum Karlrsruhe, Wissenschaftliche Berichte FZKA 6932, 2003.

23. D. D. Joseph, "Potential flow of viscous fluids: historical notes," International Journal of Multiphase Flow, vol. 32, no. 3, pp. 285–310, 2006.

24. H. B. Mahood, A. Sharif, S. A. Hossini, and R. Thorpe, "Hydrodynamics of two-phase bubbles condensation in an immiscible liquid media," In press.

25. F. Concha, "Settling velocities of particulate systems," Kona Powder and Particle Journal, vol. 27, pp. 18–37, 2009.

26. R. K. Wanchoo and G. K. Raina, "Motion of a two—phase bubble through a quiescent liquid," The Canadian Journal of Chemical Engineering, vol. 65, no. 5, pp. 716–722, 1987. ·

27. R. Olander, S. Oshmyansku, K. Nichols, and D. Werner, "Final phase testing and evaluation of the 500kWe direct contact pilot plant at East Mesa," U.S.D.O.E. Report DOE/SF/11700-T1, Arvada, Colo, USA, 1983.

28. M. Golafshani, Stability of a direct contact heat exchanger [Ph.D. thesis], University of Utah, 1984.

29. S. Sideman and G. Hirsch, "Direct contact heat transfer with change of phase: condensation of a single vapour bubbles in an immiscible liquid medium: preliminary sui," AIChE Journal, vol. 11, no. 6, pp. 1019–1025, 1965.

Chapter 8

Numerical Investigation on Double Shell-Pass Shell-and-Tube Heat Exchanger with Continuous Helical Baffles

Shui Ji, Wen-jing Du, Peng Wang, and Lin Cheng

Institute of Thermal Science & Technology, Shandong University, Jinan 250061, China

ABSTRACT

A double shell-pass shell-and-tube heat exchanger with continuous helical baffles (STHXCH) has been invented to improve the shell-side performance of STHXCH. At the same flow area, the double shell-pass STHXCH is compared with a single shell-pass STHXCH and a conventional shell-and-tube heat exchanger with segmental baffles (STHXSG) by means of numerical method. The numerical results show that the shell-side heat transfer coefficients of the novel heat exchanger are 12–17% and 14–25% higher than those of STHXSG and single shell-pass STHXCH, respectively; the shell-side pressure drop of the novel heat exchanger is slightly lower than that of STHXSG and 29–35% higher than that of single shell-pass STHXCH. Analyses of shell-side flow field show that, under the same flow rate, double shell-pass STHXCH has the largest shell-side volume average velocity and the most uniform velocity distribution of the three STHXs. The shell-side helical flow pattern of double shell-pass STHXCH is more similar to longitudinal flow than that of single shell-pass STHXCH. Its distribution of fluid mechanical energy dissipation is also uniform. The double shell-pass STHXCH might be used to replace the STHXSG in industrial applications to save energy, reduce cost, and prolong the service life.

INTRODUCTION

A variety of heat exchangers are used in industries, such as shell-and-tube heat exchangers (STHX), plate-fin heat exchangers, and fin-and-tube heat exchangers. More than 35–40% of heat exchangers are of the shell-and-tube type due to their robust geometry construction, easy maintenance, and possible upgrades [1, 2]. However, the traditional STHX with segmental baffles (STHXSG) has many disadvantages such as large back mixing, fouling, high leakage flow, and large cross-flow. Especially, segmental baffles bring on significant pressure drop across the exchanger when changing the direction of flow [3, 4]. Over the past decades, different kinds of baffles have been developed, for example, the conventional segmental baffles with different arrangements, the deflecting baffles, the overlap helical baffles, and the rod baffles [5–10].

For STHX with continuous helical baffles (STHXCH), the shell-side flow passes a periodic helical path under the action of baffles. Compared to the conventional STHXSG, STHXCH has some advantages such as reduced shell-side fouling, increased heat-transfer-rate-to-pressure-drop ratio, reduced bypass effects, and prevention from flow-induced vibration [11, 12]. Lutcha and Nemcansky [11] give two reasons accounting for the improvement of heat exchangers performance with helical baffles: firstly, a near plug flow is formed in the shell side which can increase temperature difference for heat transfer, and, secondly, the rotational flow induced by helical baffles creates a vortex which interacts with the boundary layer on the tube surface and favorably affects the heat transfer coefficient.

Some researches [13, 14] indicated that the larger the helix angle, the better shell-side comprehensive performance of STHXCH when helix angle is less than 45°. However, a large helix angle, or in other words a large helix pitch, has some adverse effects: first, the shell-side velocity becomes small under the same mass flow rate, which goes against heat transfer; second, the quantity of helical cycle is small, which means the helix flow is possibly not fully developed until it reaches the shell-side outlet; third, the unsupported span on the tube bundles is large, which is not favorable for the prevention of fluid-induced vibration in the shell side [14]. To overcome the aforementioned drawbacks of STHXCH with large helix angles, a new type of double shell-pass STHXCH is invented in this paper. At the same helix angle and shell diameter, the helix pitch and flow area of double shell-pass STHXCH become half of those of single shell-pass STHXCH. In order to validate the advantages of double shell-pass STHXCH, its performance has been compared with that of a conventional STHXSG and single shell-pass STHXCH by numerical method.

PHYSICAL MODEL

Physical models for the computational domain are depicted in Figure 1. Three models have the same heat transfer area and shell-side flow area. The geometric parameters are summarized in Table 1.

Table 1: Geometric parameters of STHXs

Item	STHXCH	STHXCH	STHXSG
Shell inside diameter (mm)	207	207	207
Tube outside diameter (mm)	19	19	19
Tube pitch (mm)	25	25	25
Length of tubes l(mm)	1500	1500	1500
Number of tubes	55	55	55
Helix angle (°)	36	20	
Helix pitch (mm)	472.5	236.7	
Baffle pitch (mm)			118.5
Flow area (m²)	7.44e-3	7.44e-3	7.44e-3
Shell pass	Double	Single	Single

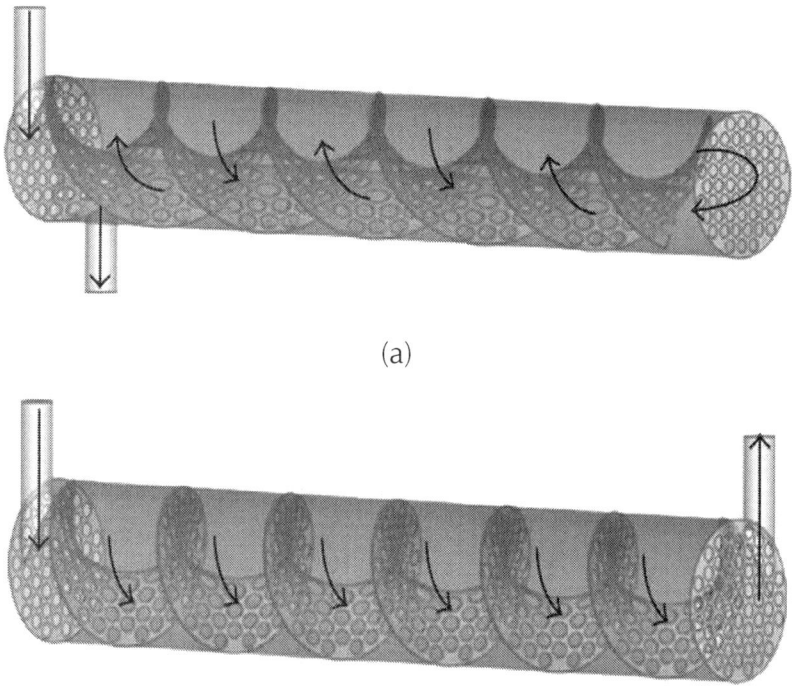

(a)

(b)

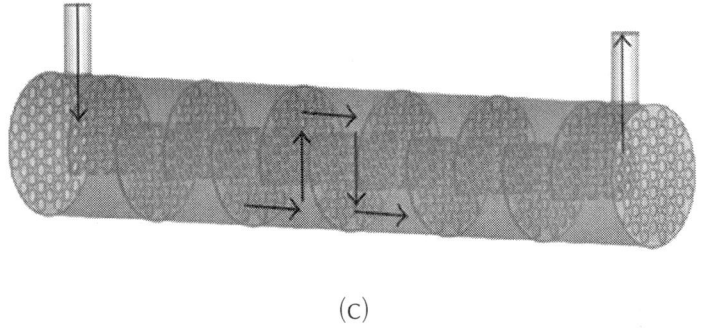

(c)

Figure 1: Physical models and flow direction.

NUMERICAL MODEL AND SIMULATION METHOD

Governing Equations

The renormalization group (RNG) $k-\varepsilon$ turbulence model is adopted because it can provide improved predictions of near-wall flows and flows with high streamline curvature [15]. The general governing equation is as follows:

$$\frac{\partial(\rho u_i \Phi)}{\partial x_i} = \frac{\partial}{\partial x_i}\left(\Gamma \frac{\partial \Phi}{\partial x_i}\right) + S.$$

(1)

The (RNG) $k-\varepsilon$ turbulence model governing equations are as follows:

$$\frac{\partial(\rho k u_i)}{\partial x_i} = \frac{\partial}{\partial x_j}\left(\alpha_k \mu_{\text{eff}} \frac{\partial k}{\partial x_j}\right) + G_k + \rho \varepsilon,$$

$$\frac{\partial(\rho \varepsilon u_i)}{\partial x_i} = \frac{\partial}{\partial x_j}\left(\alpha_\varepsilon \mu_{\text{eff}} \frac{\partial \varepsilon}{\partial x_j}\right) + \frac{C_1^*}{k}G_k - C_2\rho \frac{\varepsilon^2}{k}.$$

(2)

For continuity equation, $\Phi = 1$, generalized diffusion coefficient $\Gamma = 0$, and source term $S = 0$; for momentum equation, $\Phi = u, v, w$, generalized diffusion coefficient $\Gamma = \mu_{eff} = \mu + \mu_t$, and source term $S = -\partial p/\partial x_i + \partial/\partial x_i (\mu_{eff} (\partial u_j/\partial x_j))$; for energy equation, $\Phi = T$, generalized diffusion coefficient $\Gamma = (\mu/Pr) + (\mu_t/\sigma_T)$, and source term $S = 0$. Other relative parameters are as follows: $\mu_t = \rho C_\mu \mu (k^2/\varepsilon)$, $C_\mu = 0.0845$, $\alpha_k = \alpha_\varepsilon = 1.39$, $C_1^* = C_1 - (\eta(1-\eta/\eta_0)/(1+\beta\eta^2))$, $C_1 = 1.42$, $C_2 = 1.68$, $\eta = (2E_{ij} \cdot E_{ij})^{1/2}(k/\varepsilon)$, $E_{ij} = 1/2((\partial u_i/\partial x_j) + (\partial u_j/\partial x_i))$, $\eta_0 = 4.377$, and $\beta = 0.012$, among which μ is viscosity coefficient, Pr is Prandtl number, σ_T is turbulent Prandtl number, p is pressure, T is temperature, and u, v, and w are velocity components.

Basic Assumptions and Boundary Conditions

To simplify the numerical simulation while still keeping the basic characteristics of the process, the following assumptions are made: (1) both the fluid flow and heat transfer processes are turbulent and in steady state; (2) the leakage flow between the tube and the baffle and that between the baffle and the shell are neglected; (3) effects of gravity and buoyancy forces are neglected; (4) the tube wall temperatures are kept constant in the whole shell side; (5) the heat exchanger is well insulated hence, the heat loss to the environment is totally neglected; (6) and the working fluid is heat-transfer oil. Its viscosity changes in a large extent with the variation of temperature, and hence a quadratic function is fitted between viscosity and temperature, and the other physical properties are regarded as constant.

The shell-side inlet is set as velocity inlet with a prescribed flow rate and temperature ($T_{in}=80°C$). The outlet port is set as a pressure boundary condition, which means that a static pressure and a proper backflow are specified. The temperature of the tube walls is set as a constant of 40°C and the other surfaces are set as nonslip, adiabatic, and impermeable.

Mesh Generation and Numerical Method

Due to the complicated structure of STHXCH, the computational domain is meshed with unstructured tetrahedral and pyramidal elements which are generated by ICEM 12.0. Mesh adoption is also used for refining and coarsening local mesh according to gradient of variables. In order to ensure the accuracy of numerical results, a careful test for the mesh in-dependence of the numerical solutions was conducted. In the test, three different mesh systems with 9.8 million, 13.6 million, and 17.8 million elements are adopted for calculation of the whole heat exchanger, and the difference in the overall pressure drop and the shell-side heat transfer coefficient between the last two mesh systems is less than 2%. The local meshes and velocity distributions are shown in Figure 2.

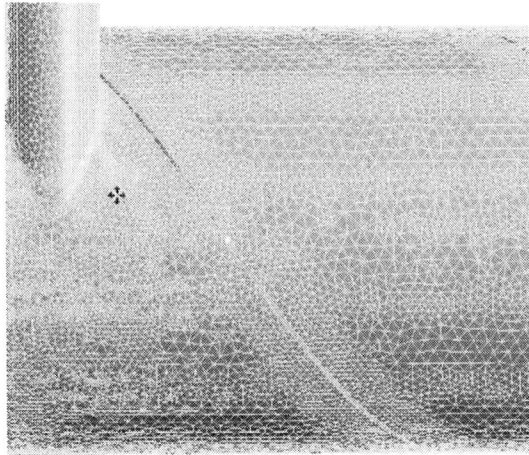

Figure 2: Local meshes of computational model.

The commercial code ANSYS CFX 12.0 is adopted to simulate the flow and heat transfer in the computational model. The governing equations are discredited by the finite volume method. The convergence criterion is that the flow field and mass residual should be less than 10^{-6} for the energy residual less than 10^{-7} for the energy equation. A parallel computation is performed on four

DELL workstations with two Quad-Core CPUs and 16 GB memory each by using CFX, and every simulation case takes approximately 36 h to get converged solutions.

Validation of Numerical Method

Validation of the numerical method was made using experimental results from the literature [16]. Figure 3 provides the comparisons between experimental date and simulation results using the numerical method in this paper. It can be observed that for both fluid pressure drop and heat transfer their variation trends with mass flow rate in good agreement. Quantitatively, the difference in pressure drop is 6.5%~17.8% and the difference in the heat transfer coefficients ranges 2.7%~5.1%. Obviously the numerical method in this paper is reliable and applicable.

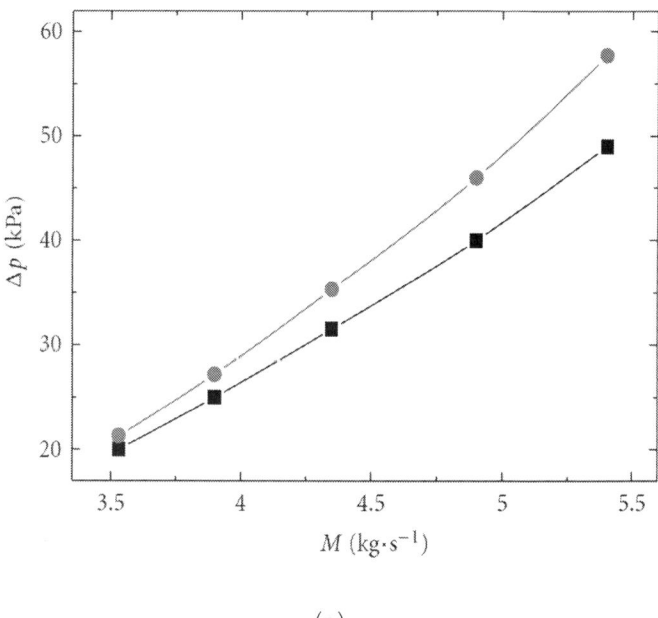

(a)

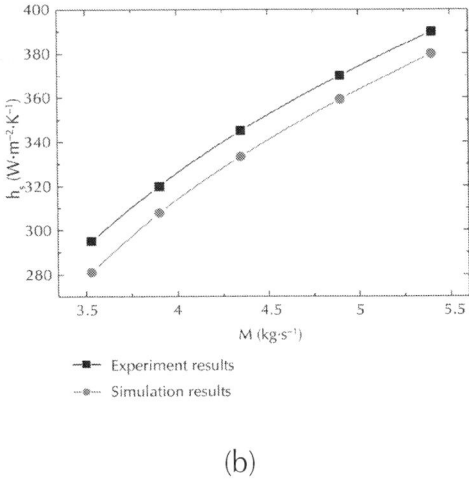

(b)

Figure 3: Comparison experimental results and simulation results in shell side.

RESULTS AND DISCUSSION

Heat Transfer and Pressure Drop

Heat exchanger rate Q_s of shell-side fluid is as follows:

$$Q_s = M_s \times c_{ps} \times \left(t_{s,\text{in}} - t_{s,\text{out}}\right). \qquad (3)$$

The shell-side heat transfer coefficient h_s is equal to

$$h_s = \frac{Q_s}{A_o \cdot \Delta t_m},$$

$$A_o = N_t \cdot \pi d_o l,$$

$$\Delta t_m = \frac{\Delta t_{\max} - \Delta t_{\min}}{\ln(\Delta t_{\max}/\Delta t_{\min})},$$

$$\Delta t_{\max} = t_{s,\text{in}} - t_w,$$

$$\Delta t_{\min} = t_{s,\text{out}} - t_w, \qquad (4)$$

Where A_o is the heat exchange area based on the outer diameter of tube d_o, N_t is the number of tubes, l is the effective length of tubes, t_w is the temperature of tube walls and the subscripts s and t refer to shell side and tube side, respectively.

The variation trends of shell-side heat transfer and pressure drop with mass flow rate are shown in Figures 4 and 5. It can be seen that, at the same mass flow rate and flow area, the shell-side heat transfer coefficients and heat transfer rate of double shell-pass STHXCH are 12–17% and 14–25% higher than those of STHXSG and single shell-pass STHXCH, respectively; the shell-side pressure drop of double shell-pass STHXCH is slightly lower than that of STHXSG and 29–35% higher than that of single shell-pass STHXCH.

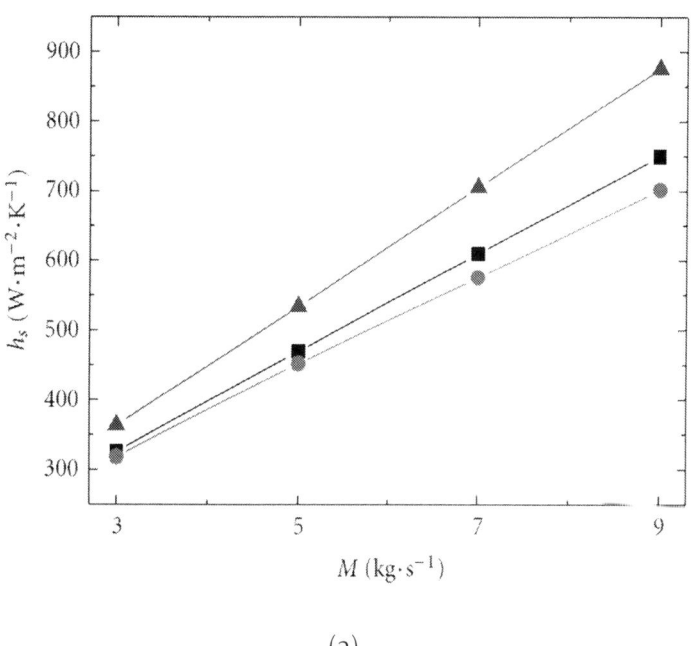

(a)

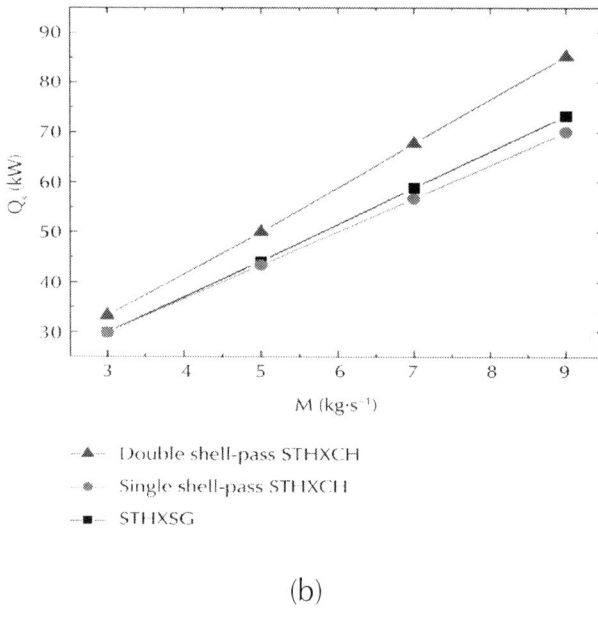

(b)

Figure 4: Shell-side heat transfer performance.

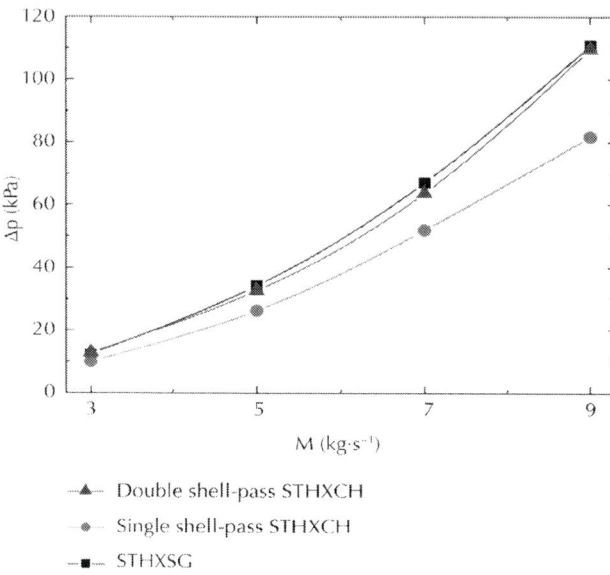

Figure 5: Shell-side pressure drop versus flow rate.

Flow Field Analysis

In Figure 6 the shell-side volume average velocity data are presented. It can be clearly observed that the shell-side volume average velocity of double shell-pass STHXCH is much higher than those of the other two types of STHX in spite of the same flow rate and flow area. That is one of the reasons why double shell-pass STHXCH has the best heat transfer performance.

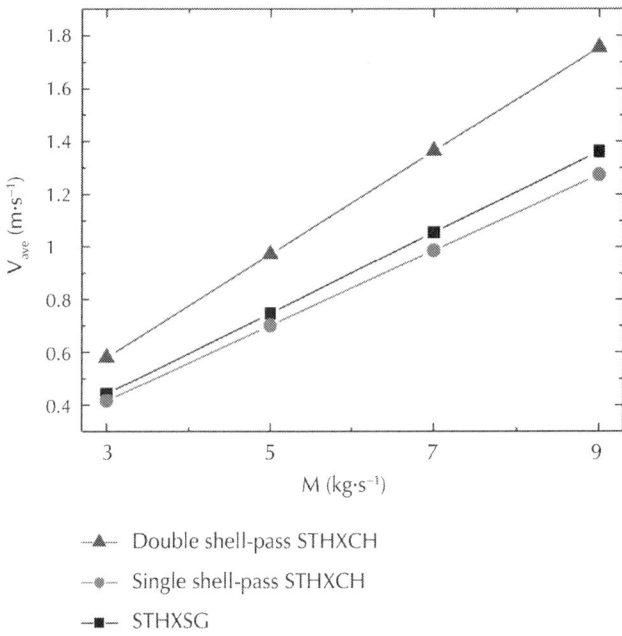

Figure 6: Shell side volume average velocity versus flow rate.

The velocity distributions of the three STHXs are shown in Figure 7 with same mass flow rate. It can be found that there is nearly no back flow regions and dead zones existed in the shell pass of STHXCHs. The velocity distribution of double shell-pass STHXCH is the most uniform of three STHXs.

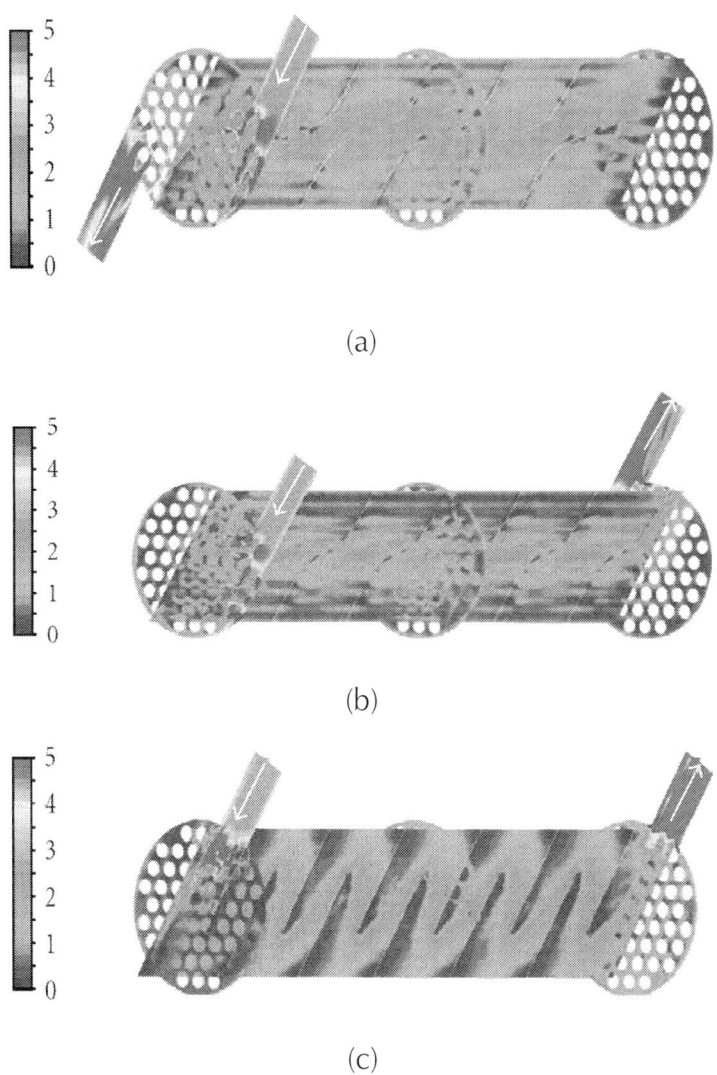

(a)

(b)

(c)

Figure 7: Velocity distributions (M=7kg·s⁻¹) (unit: m·s⁻¹).

The local velocity vector distributions on the axial sections of shell are shown in Figure 8. For both STHXCH, the shell-side fluids pass through the tube bundles basically in a helical pattern and rush the heat exchange tubes with an inclination angle. On the one hand, helical flow avoids abrupt turns of flow. On the other hand, it changes the cross-section shape of tube in flow direction into ellipse. Therefore, it can reduce the pressure drop in shell side and the vibration of tube bundle. It also can be found that, in each shell-pass of double shell-pass STHXCH, the flow directions are opposite totally. In addition, the angle between flow direction and axis of tube of double shell-pass STHXCH is much smaller than that of single shell-pass STHXCH. It means that the double shell-pass STHXCH is more similar to the longitudinal flow heat exchanger than the single shell-pass STHXCH. For STHXSG, the shell-side fluid passes through the tube bundles in a dramatic zigzag pattern, which causes large flow resistance and high risk of vibration failure on tube bundle. Furthermore, obvious dead zones are formed at the corners between baffles and shell wall. Flow stagnation in dead zones goes against heat transfer and increases fouling resistance. Therefore, the shell-side pressure drop of double shell-pass STHXCH angles is slightly lower than that of STHXSG.

(a)

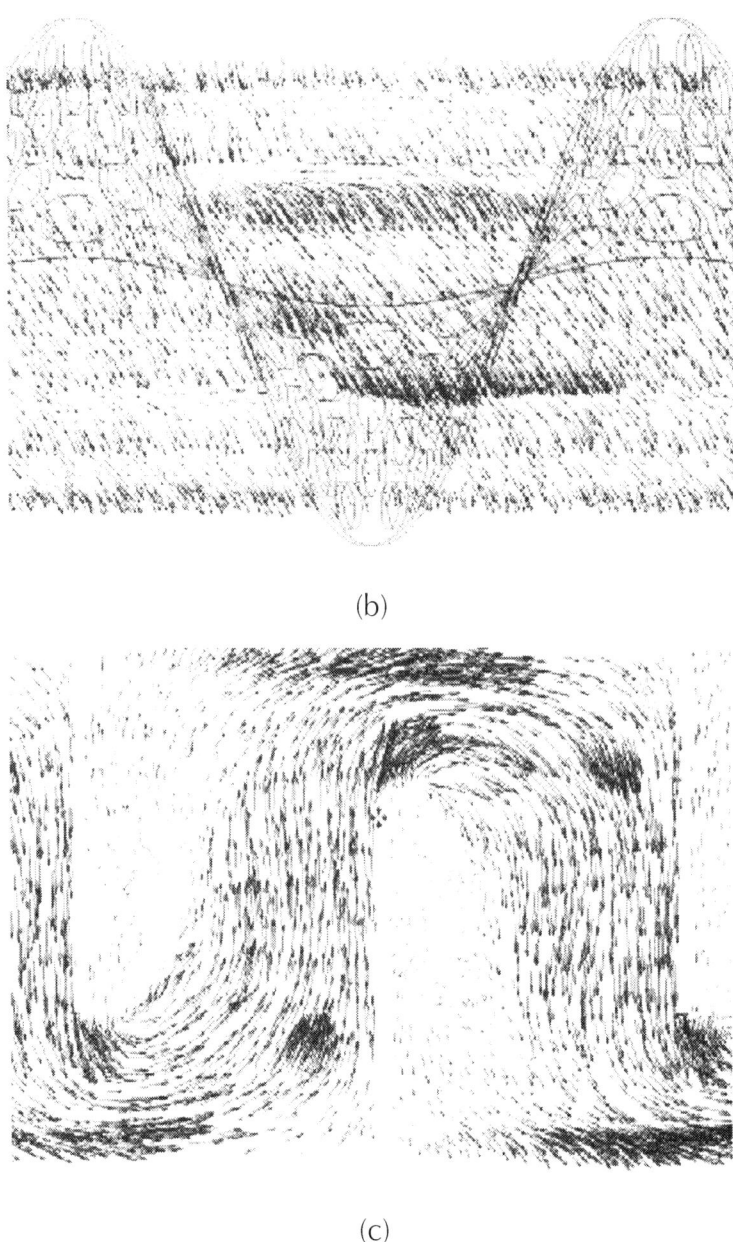

(b)

(c)

Figure 8: Local velocity vector distributions on axial sections of shell (M=7kg/s).

Mechanical Energy Dissipation

Viscousness of the fluid leads to internal friction when the fluid flows. It is the fundamental reason of the generation of flow resistance. Fixed walls or another type of solid surfaces provide the conditions for the production of flow resistance. The final effect of flow resistance is the dissipation of fluid mechanical energy. Within a (RNG) $k-\varepsilon$ approach for turbulent and steady flow, the local mechanical energy dissipation rate E_D of fluid elements is formed by two parts. They are

$$E_D = E_{\overline{D}} + E_{D'}.\tag{5}$$

The first term $E_{\overline{D}}$ is equal to viscous dissipation rate Φ, which is calculated by local time-averaging velocity and fluid dynamic viscosity. The second term $E_{D'}$ can be calculated by

$$E_{D'} = \rho \cdot \varepsilon \tag{6}$$

As shown in [17]. Here ε is the local turbulent dissipation rate, calculated with a $k-\varepsilon$ turbulence model. The E_D distributions of the three STHXs are shown in Figure 9. It can be seen that, in flow full-developed region, the E_D distribution of double shell-pass STHXCH is the most uniform of the three STHXs. For single shell-pass STHXCH, the fluid mechanical energy dissipation concentrated in the central region of the shell. For STHXSG, dissipation of fluid mechanical energy concentrated in cross-flow tube banks region of the shell.

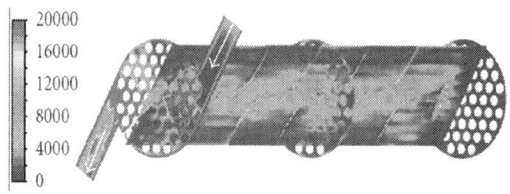

(a)

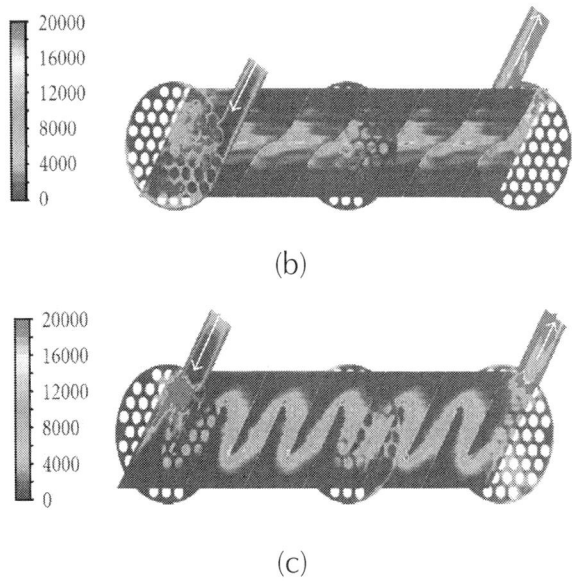

Figure 9: Mechanical energy dissipation rate E_D distributions (M=7kg·s⁻¹) (unit: W·m⁻³).

Comprehensive Performance

Figure 10 provides the comparison of shell-side heat transfer coefficient and heat transfer rate among the three types of STHX within the range of pressure drop tested.

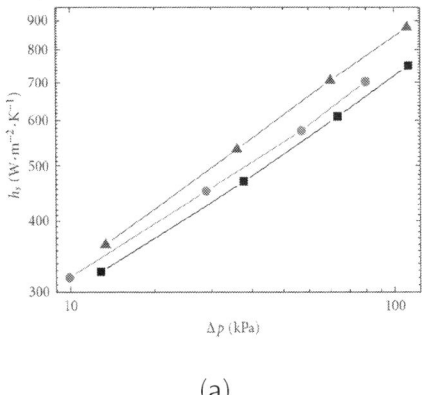

(a)

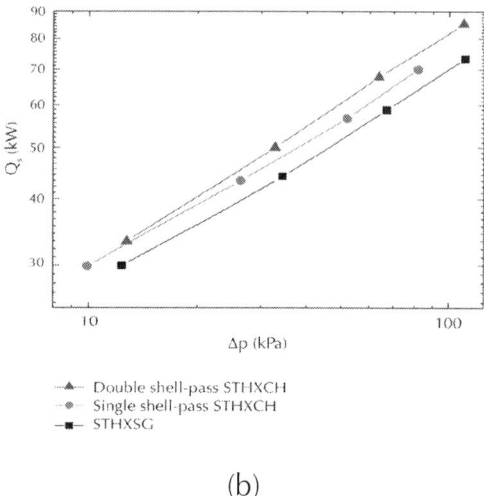

(b)

Figure 10: Shell-side comprehensive performance.

It can be found from Figure 10 that under the same pressure drop, the shell-side heat transfer coefficient of double shell-pass STHXCH is 11–18% and 5–13% higher than that of single shell-pass STHXCH and STHXSG, respectively. Then, the double shell-pass STHXCH has the best shell-side comprehensive performance and the STHXSG has the worst shell-side comprehensive performance. The double shell-pass STHXCH might be used to replace the STHXSG in industrial applications to save energy, reduce cost, and prolong the service life.

CONCLUSIONS

In this paper, a novel double shell-pass STHXCH is investigated with numerical method and compared with a single shell-pass STHXCH and a STHXSG. Three models have the same shell-side flow area. The conclusions are summarized as follows: (1) under the same flow rate, the shell side heat transfer coefficient of double shell-pass STHXCH is 14–25% and 12–17% higher than that of STHXSG and single shell-pass STHXCH, respectively; the shell-side pressure drop of double shell-pass STHXCH is 29–35% higher than that of single shell-pass STHXCH and slightly lower than that of STHXSG. (2) Under the same shell side pressure drop, the shell-side volume average velocity of double shell-pass STHXCH is the highest and the velocity distribution is the most uniform. (3) In flow full-developed region of double shell-pass STHXCH, the distribution of fluid mechanical energy dissipation rate is uniform in the three STHXs. (4) under the same pressure drop, double shell-pass STHXCH has the best heat transfer performance.

ACKNOWLEDGMENTS

The authors would like to acknowledge the financial support from the National Basic Research Program of China (973 Program) (no. 2007CB206900).

REFERENCES

1. K. J. Bell, "Heat exchanger design for the process industries," Journal of Heat Transfer, vol. 126, no. 6, pp. 877–885, 2004.

2. B. I. Master, K. S. Chunangad, A. J. Boxma, D. Kral, and P. Stehlík, "Most frequently used heat exchangers from pioneering research to worldwide applications," Heat Transfer Engineering, vol. 27, no. 6, pp. 4–11, 2006.

3. H. Li and V. Kottke, "Effect of the leakage on pressure drop and local heat transfer in shell-and-tube heat exchangers for staggered tube arrangement," International Journal of Heat and Mass Transfer, vol. 41, no. 2, pp. 425–433, 1998.

4. H. Li and V. Kottke, "Visualization and determination of local heat transfer coefficients in shell-and-tube heat exchangers for staggered tube arrangement by mass transfer measurements," Experimental Thermal and Fluid Science, vol. 17, no. 3, pp. 210–216, 1998.

5. P. Stehlík and V. V. Wadekar, "Different strategies to improve industrial heat exchange," Heat Transfer Engineering, vol. 23, no. 6, pp. 36–48, 2002.

6.   H. Li and V. Kottke, "Effect of baffle spacing on pressure drop and local heat transfer in shell-and-tube heat exchangers for staggered tube arrangement," International Journal of Heat and Mass Transfer, vol. 41, no. 10, pp. 1303–1311, 1998.

7.   R. L. Webb and N. H. Kim, Principles of Enhanced Heat Transfer, Taylor & Francis, Boca Raton, Fla, USA, 2005.

8.   E. A. Vasil›Tsov, "Turbulence of flow in mixers with deflecting baffles," Chemical and Petroleum Engineering, vol. 24, no. 3-4, pp. 111–115, 1988.

9.   C. C. Gentry, "RODbaffle heat exchanger technology," Chemical Engineering Progress, vol. 86, no. 7, pp. 48–57, 1990.

10.   Q. W. Dong, Y. Q. Wang, and M. S. Liu, "Numerical and experimental investigation of shellside characteristics for RODbaffle heat exchanger," Applied Thermal Engineering, vol. 28, no. 7, pp. 651–660, 2008.

11.   J. Lutcha and J. Nemcansky, "Performance improvement of tubular heat exchangers by helical baffles,"Chemical Engineering Research and Design, vol. 68, no. 3, pp. 263–270, 1990. ·

12.   Q. Wang, G. Chen, Q. Chen, and M. Zeng, "Review of Improvements on shell-and-tube heat exchangers with helical baffles," Heat Transfer Engineering, vol. 31, no. 10, pp. 836–853, 2010.

13.   Y. G. Lei, Y. L. He, R. Li, and Y. F. Gao, "Effects of baffle inclination angle on flow and heat transfer of a heat exchanger with helical baffles," Chemical Engineering and Processing: Process Intensification, vol. 47, no. 12, pp. 2336–2345, 2008.

14.   S. Ji, W. J. Du, and L. Cheng, "Numerical investigation on heat transfer and flow properties in shell-side of heat exchanger with continuous helical baffles," Proceedings of the Chinese Society of Electrical Engineering, vol. 29, no. 32, pp. 66–70, and 2009.

15. V. Yakhot and L. M. Smith, "The renormalization group, the ε-expansion and derivation of turbulence models," Journal of Scientific Computing, vol. 7, no. 1, pp. 35–61, 1992.

16. H. U. Yan, Numerical Simulation of Shell-and-Tube Heat Exchanger, Harbin Institute of Technology, Harbin, China, 2007.

17. F. Kock and H. Herwig, "Local entropy production in turbulent shear flows: a high-Reynolds number model with wall functions," International Journal of Heat and Mass Transfer, vol. 47, no. 10-11, pp. 2205–2215, 2004.

Numerical Investigation of Al$_2$O$_3$/Water Nanofluid Laminar Convective Heat Transfer through Triangular Ducts

Saeed Zeinali Heris, Seyyed Hossein Noie, Elham Talaii, and Javad Sargolzaei

Chemical Engineering Department, Faculty of Engineering, Ferdowsi University of Mashhad, Mashhad, Iran

ABSTRACT

In this article, laminar flow-forced convective heat transfer of Al$_2$O$_3$/water nanofluid in a triangular duct under constant wall temperature condition is investigated numerically. In this investigation, the effects of parameters, such as nanoparticles diameter, concentration, and Reynolds number on the enhancement of nanofluids heat transfer is studied. Besides, the comparison between nanofluid and pure fluid heat transfer is achieved in this article. Sometimes, because of pressure drop limitations, the need for non-circular ducts arises in many heat transfer applications. The low heat transfer rate of non-circular ducts is one the limitations of these systems, and utilization of nanofluid instead of pure fluid because of its potential to increase heat transfer of system can compensate this problem. In this article, for considering the presence of nanoparticl: es, the dispersion model is used. Numerical results represent an enhancement of heat transfer of fluid associated with changing to the suspension of nanometer-sized particles in the triangular duct. The results of the present model indicate that the nanofluid Nusselt number increases with increasing concentration of nanoparticles and decreasing diameter. Also, the enhancement of the fluid heat transfer becomes better at high Re in laminar flow with the addition of nanoparticles.

INTRODUCTION

The increase of heat transfer coefficient is one of the most important technical aims for industry and researches. Also, the decrease in the pressure drop for systems that generate high fluid pressure drop is very noticeable. The aim, therefore, for achieving the optimization of heat exchangers must be always to increase the heat transfer, and simultaneously minimize the increase in the pressure drop [1]. Increased efforts are being directed to produce heat exchangers with higher efficiency to achieve savings of energy, material, and labor [2]. Improvements in heat transfer augmentation depend on performance and manufacturing cost [3]. Consequently, there is an increased need for utilization of a variety of duct geometries for heat transfer applications with forced convection and internal flow [2]. Because of the size and volume constraints in applications, such as aerospace, nuclear, biomedical engineering, and electronics, the utilization of non-circular flow passage geometries may be required, particularly, in respect of compact heat exchangers [2]. Consequently, duct with non-circular cross section (triangular) is used in this study because of its low pressure drop, but it causes decrement of heat transfer. Therefore, for compensating this decrement, nanofluid, instead of pure fluid, was used in this study because of the former's potential to increase the heat transfer of the system. Nanofluids are created by dispersing nanometer-sized particles (<100 nm) in a base fluid such as water, ethylene-glycol, or propylene-glycol [4].

Choi [5] was the first person to have created fluids containing suspension of nanometer-sized particles which are called the nanofluids and disclosed their significant thermal properties through the measurement of the convective heat transfer coefficient of those fluids. Various benefits of the application of nanofluids, such as improved heat transfer, size reduction of the heat transfer system, minimal clogging, microchannel cooling, and miniaturization of systems, were achieved in his study. Since then investigations have been continued at three phases as described below:

- Conductive heat transfer was investigated in studies by many researchers, e.g., Lee et al. [6], which include measurement of conductive heat transfer coefficients of Al_2O_3/water, Al_2O_3/ethylene-glycol, CuO/water, and CuO/ethylene-glycol nanofluids.

- Convective heat transfer was also studied in some published articles. For example, Pak and Cho [7] investigated convective heat transfer in the turbulent flow regime using Al_2O_3/water and TiO_2/water nanofluids, and found that the Nusselt number of the nanofluids increased with increasing volume fraction of the suspended nanoparticles, and the increasing Reynolds number. Lee and Choi [8] studied convective heat transfer of laminar flows of an unspecified nanofluid in microchannels, and observed a reduction in thermal resistance by a factor of 2. Nanofluids were also observed to have the ability to dissipate a heat power three times more than pure water could do. Xuan and Li [9] measured convective heat transfer coefficient of Cu/water nanofluids, and found substantial heat transfer enhancement. For a given Reynolds number, heat transfer coefficient of nanofluids containing 1% volume Cu nanoparticles was shown to be approximately 12% higher than that of pure water.

Zienali Heris et al. [10-12] investigated the convective heat transfer of Al_2O_3/water and CuO/water nanofluids in circular tubes, and observed that the heat transfer coefficient was enhanced by increasing the concentration of nanoparticles in the nanofluids. However, the 20-nm Al_2O_3 nanoparticles showed an improved heat transfer performance compared with the 50-nm CuO nanoparticles, especially at high concentrations.

Maiga et al. [13] studied numerically the heat transfer enhancement in turbulent tube flow using Al_2O_3 nanoparticles suspension. Their results showed that the inclusion of nanoparticles into the base fluid produced an augmentation of the heat transfer coefficient which has been found to increase appreciably with an increase in the concentration of particles.

Akbari and Behzadmehr [14] investigated the developing laminar-mixed convection flow of a nanofluid consisting of Al_2O_3/water in a horizontal tube and hypothesized that the nanoparticles' concentration did not have any significant effect on the secondary flow pattern and the axial velocity.

Das and Ohal [15] studied numerically the behavior of nanofluids inside a partially heated and partially cooled square cavity to gain insight into heat transfer and flow processes induced by a nanofluid. Ben Mansour et al. [16] investigated the conjugate problem of developing laminar-mixed convection flow and heat transfer of Al_2O_3/water nanofluid inside an inclined tube subjected to a uniform wall heat flux. The Cu/water nanofluid-forced convective heat transfer performance in a circular tube was experimentally investigated by Zeinali Heris et al. [17]. Based on their experimental results, they observed that the heat transfer coefficient was influenced by Peclet number, as well as by Cu nanoparticles' volume concentrations. They also stated that there was an optimum concentration for Cu nanoparticles in water, in which improved enhancement for heat transfer can be found [17].

- Boiling heat transfer: The boiling process of nanofluids was investigated experimentally by several researchers [18-21].

Das et al. [18] observed the nanofluids' boiling performance deterioration. Soltani et al. [19], through experimental measurements of boiling heat transfer characteristics of Al_2O_3/water and SnO_2/water Newtonian nanofluids, showed that nanofluids possess noticeably higher boiling heat transfer coefficients than those of the base fluid.

Bang and Chang [21] studied boiling heat transfer characteristics of nanofluids with alumina nanoparticles suspended in water. They found that the addition of alumina nanoparticles caused a decrease of pool nucleate boiling heat transfer.

Noie et al. [22] investigated heat transfer enhancement using Al_2O_3/water nanofluids in a two-phase closed thermosyphon. Experimental results showed that for different input powers, the efficiency of the TPCT increases up to 14.7% when Al_2O_3/water nanofluid was used instead of pure water. The comparison between heat transfer enhancements using metallic and oxide nanoparticles was done by Hamed Mosavian et al. [23] whose results indicated the enhancement of heat transfer with increasing nanoparticles. Based on their experimental results, metallic nanoparticles showed better enhancement of heat transfer coefficient in comparison with oxide particles.

There are many passive cases regarding nanofluids that are still unrecognized. The most of the investigations were regarding heat transfer in circular ducts, and there is no report regarding ducts with a triangular cross section which causes lower pressure drop than the other forms of ducts. Different criteria for selecting and optimizing the heat exchanger passage geometries were outlined by Bergles [24]. Kays and London [25] showed that a compact heat exchanger, with a triangular cross-sectional internal flow passage, had a high ratio of heat transfer area to flow-passage volume. As far back as the late 1950s and early 1960s, Eckert et al. [26], Sparrow [27], Sparrow and Haji-Sheikh [28], and Schmidt and Newell [29] used approximate solution methods to study the pressure drop and convective heat transfer in fully developed laminar flow in ducts with cross sections having an equilateral or isosceles triangular section. Shah [30] and Shah and London [31] studied the heat transfer characteristics of laminar flow in a wide variety of channel shapes, including equilateral triangular, equilateral triangular with rounded corners, isosceles triangular, right triangular, and arbitrary triangular cross-sectional ducts, for an extensive range of thermal boundary conditions. Recently, Zhang [32] has reported Nusselt numbers for laminar hydrodynamically fully developed and thermally developing flow for a uniform wall temperature condition in isosceles triangular ducts with apex angles ranging from 30° to 120°. Gupta et al. [33] studied fully developed laminar flow and heat transfer in equilateral triangular cross-sectional ducts following serpentine and trapezoidal path. Kuznetsov et al. [34] studied the effects of thermal dispersion and turbulence in forced convection in a composite parallel-plate channel and stated: "Although the flow in the porous region remains laminar, thermal dispersion may have a dramatic impact on heat transfer in the channel."

The aim of this article is to study the laminar-forced convective flow heat transfer of nanofluid in a triangular duct with constant wall temperature using the dispersion model.

Mathematical Modeling

Laminar flow-forced convection of Al_2O_3/water nanofluid in a triangular duct is studied numerically. The duct configurations and coordinate system are shown in Figure 1.

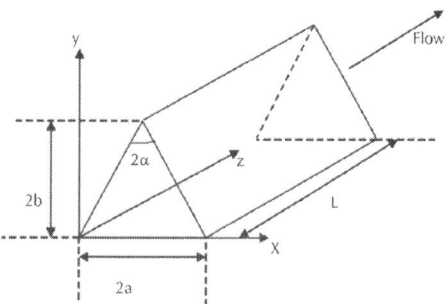

Figure 1: Geometry of a triangular duct.

The assumptions of the model presented in this article can be summarized as follows:

- Fully developed, steady-state, laminar flow.
- Constant wall temperature.
- Neglecting the axial diffusion terms in the equations of conservation of momentum and energy.
- Neglecting x, y direction convective terms.
- Equilateral triangular duct considered for investigation.
- Neglecting viscous dissipations in Cartesian coordinates

For the hydro dynamically developed and thermally developing flow, there is only one nonzero component of velocity (u), and the constitutive equations of motion reduce to a single nonlinear partial differential equation of the form:

$$\frac{\partial}{\partial x}\left(\mu \frac{\partial u}{\partial x}\right) + \frac{\partial}{\partial y}\left(\mu \frac{\partial u}{\partial y}\right) = \frac{dp}{dz}$$

(1)

The dimensionless parameters are defined as follows:

$$Z = \frac{z}{2a}$$

(2)

$$Y = \frac{y}{2b}$$

(3)

$$X = \frac{x}{2a}$$

(4)

$$\theta = \frac{T - T_w}{T_i - T_w}$$

(5)

$$u^* = \frac{-\mu u}{\left(\dfrac{\mathrm{d}p}{\mathrm{d}Z} \right) D_h^2}$$

(6)

The dimensionless momentum equation can be written as [32]

$$\frac{\partial^2 u^*}{\partial X^2} + \left(\frac{a}{b} \right)^2 \frac{\partial^2 u^*}{\partial Y^2} + \frac{4a^2}{D_h^2} = 0$$

(7)

Consequently, the dimensionless velocity profile for equilateral triangular duct defining $U = \dfrac{u}{u_m}$ is calculated as [31]

$$U = \frac{15}{b^2} \left[-b^2 Y^3 + 3a^2 YX + \left(b^2 Y^2 + a^2 X^2 \right) - \left(\frac{4}{27} \right) b^2 \right]$$

(8)

On the basis of the above assumptions, the energy equation for constant property flow is defined as

$$\frac{k_{eff}}{\rho Cp} \left(\frac{\partial}{\partial x} \left(\frac{\partial T}{\partial x} \right) + \frac{\partial}{\partial y} \left(\frac{\partial T}{\partial y} \right) \right) = u \frac{\mathrm{d}T}{\mathrm{d}z}$$

(9)

Where k_{eff} is the effective thermal conductivity of the nanofluid. There are different approaches that are discussed in the literature for investigating the heat transfer of nanofluids. In the first approach (homogeneous model), the flow and energy equations of the base fluid are not affected by the presence of the nanoparticles. Under this assumption, both the fluid phase and solid nanoparticles flow at the same velocity and are in thermal equilibrium [35,36].

In the second approach, for the contribution of hydrodynamic dispersion and irregular movement of the nanoparticles, modified homogeneous model or dispersion model is adopted [35,36].

In this investigation, the dispersion model, in which the effect of random movement of nanoparticles inside the liquid is considered as excess terms in the heat transfer equation, is solved. The effective thermal conductivity of the nanofluid may take the following form [35]:

$$k_{eff} = k_{nf} + k_d$$

(10)

where k_d is the dispersion thermal conductivity. With respect to the similarity between diffusion in porous media and nanofluid flow, the following formula has been proposed to calculate k_d [37-39]:

$$k_d = c(\rho Cp)_{nf} u_m v d_p a$$

(11)

where (c) is an unknown constant, and should be determined by matching experimental data. It depends on the diameter of the nanoparticles and flowing surface geometry. The comparison of the measured values of the nanofluid thermal conductivity with the calculated values from the proposed models indicates that the thermal dispersion is the main mechanism for enhancing fluid thermal conductivity inside channels filled with nanofluids under convective conditions. In fact, Equation 11 is a first approximation considering the dispersive effects of nanoparticles on the thermal conductivity of the nanofluid flowing through channels. According to the study of Khaled and Vafai [40] in which heat transfer of nanofluid flow in a channel was investigated, the range of the value of c was chosen to be from 0 to 0.4. Comparing this study (triangular duct) with the channel flow in the Khaled and Vafai's investigation, the value of c = 0.3 is used in this study. In order to examine the exact value of constant (c), further experimental and numerical investigations are needed.

Finally, the energy equation for laminar flow in an equilateral triangular duct is

$$2aU\frac{\partial\theta}{\partial Z} = \left[\frac{k_{nf}}{(\rho Cp)_{nf}u_m} + \frac{c(\rho Cp)_{nf}vd_p a}{(\rho Cp)_{nf}}\right]\frac{\partial^2\theta}{\partial X^2} + \left(\frac{a}{b}\right)^2\left[\frac{k_{nf}}{(\rho Cp)_{nf}u_m} + \frac{c(\rho Cp)_{nf}vd_p a}{(\rho Cp)_{nf}}\right]\frac{\partial^2\theta}{\partial Y^2}$$

(12)

where Peclet number can be used for simplifying the equation

$$Pe_{nf} = \frac{2au_m(\rho Cp)_{nf}}{k_{nf}}$$

(13)

Consequently, the temperature distribution equation takes the form:

$$\theta_b = \frac{\iint u^* \theta \, dA}{\iint u^* \, dA}$$

(14)

Dimensionless bulk temperature is defined in the following form [40]:

An energy balance in a control volume in the duct will give the equation for the estimation of the local Nusselt number as follows:

$$Nu = \frac{-\left(\dfrac{\partial \theta}{\partial X}\right)_{X=1}}{\theta_b}$$

(15)

Thermophysical Properties of Nanofluids

The thermophysical properties of nanofluid in the Equations 13-14 were calculated from nanoparticles and water properties using the following correlation at the mean bulk temperature[5,35,41]:

$$(\rho Cp)_{nf} = (1 - v)(\rho Cp)_f + v(\rho Cp)_s$$

(16)

Thermal conductivity is the most important parameter indicating the enhancement potential of the nanofluids. Based on the studies carried out to evaluate the thermal conductivity of the nanofluids[5,6,42-45], the theoretical models cannot predict the thermal conductivity of nanofluids. In the absence of experimental data, Yu and Choi's correlation [44] used for the determination of the nanofluid's effective thermal conductivity is as follows:

$$k_{nf} = \left[\frac{k_s + 2k_w + 2(k_s - k_w)(1 + \beta)^3 v}{k_s + 2k_w - (k_s - k_w)(1 + \beta)^3 v} \right] k_w$$

(17)

where β is the ratio of the nanolayer thickness to the original particle radius, and $\beta = 0.1$ was used to calculate the nanofluid's effective thermal conductivity.

Validation of the Simulation

In this article, the finite difference method is used for numerical solution. The discretization in the physical space (x, y) is performed by dividing the flow domain in equal triangular elements. The grid is constructed by drawing inside the triangular cross section three groups of parallel lines. The lines of each group are equally distanced and parallel to one of the three sides of the triangle.

The benefit of such a discretization is obvious. Since the boundaries of the computational domain are identical to the boundaries of the triangular cross sections of the channel, this method provides good accuracy in the numerical solution [45,46]. Hexagonal computational cell and finite differencing along a characteristic is shown in Figure 2.

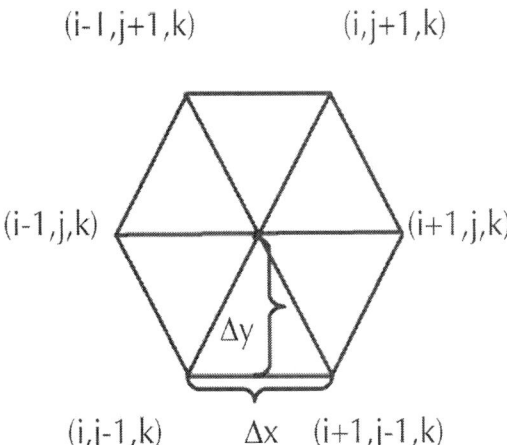

Figure 2: Hexagonal computational cell and finite differencing along a characteristic.

Equation 14 was discretized using central differencing for

$\left(\dfrac{\partial^2\theta}{\partial Y^2}\right)$ and $\left(\dfrac{\partial^2\theta}{\partial X^2}\right)$ and backward differencing for $\left(\dfrac{\partial\theta}{\partial Z}\right)$

Boundary conditions in dimensionless form read as follows:

$\theta(i,j,1)=1$ (19)

$\theta(1,j,k)=0$ (20)

$\theta(i,1,k)=0$ (21)

$\theta(i,m,k)=0$ (22)

$$\frac{\partial\theta}{\partial x}\left(\frac{m+1}{2},j,k\right)=0$$

(23)

where m is gride number in x and y directions. The grid used in the present analysis is $560 \times 560 \times 100$ nonuniform one with highly packed grid points in the vicinity of the tube wall and especially in the entrance region (560 in x, y direction; 100 in z direction). In order to ensure grid independence, the solution is tested for the $800 \times 800 \times 180$ and the $500 \times 500 \times 90$ configurations, with all of the latter giving similar values. Therefore, $560 \times 560 \times 100$ configuration was accepted as the optimal grid size.

The duct is an equilateral triangle with a length of each side being 100 cm. The flow is laminar with the Reynolds number ranging from 100 to 2100. Because of the absence of experimental data for nanofluid in triangular ducts, in order to validate the computational model, the numerical results were compared with the theoretical data available for the conventional fluids in triangular duct as proposed by Shah and London [31]. Figure 3 displays the comparison of Nusselt number values computed by Shah and London [31] with the computed values from the present simulations.

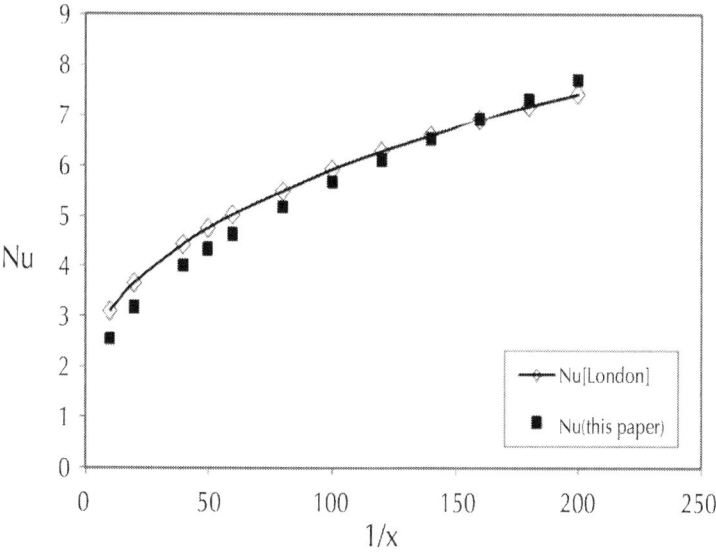

Figure 3: Comparison between model predictions and results defined by Shah and London [31].

RESULTS AND DISCUSSION

In this section, the effect of nanoparticle's diameter, nanoparticle's concentration, and Reynolds number on heat transfer performance of Al_2O_3/water nanofluid is investigated.

Figure 4 shows the average Nusselt numbers versus Re for Al_2O_3/ water nanofluid (with 1.0% volume concentration of 10-nm Al_2O_3 nanoparticles) and pure water. As shown in Figure 4, the slope of *Nu* versus *Re* is greater for Al_2O_3/water compared with pure water, which means a considerable enhancement of heat transfer due the addition of nanoparticles to the base fluid. For example, at *Re* = 1500, Nusselt number of water is increased from 3.47 to 4.22 with the addition of Al_2O_3 nanoparticles. It is known that the addition of Al_2O_3 nanoparticles will increase the thermal conductivity of the working fluid and hence the heat transfer capability [5-12].

Besides, the nanoparticles with dispersion effect and Brownian motion hit the tube wall and absorb heat, and then mix back with the bulk of the fluid to cause a better heat transfer. The presence of nanoparticles inside the fluid causes the collision between the heating surface and the particles, thereby producing higher heat transfer coefficients. This means that the addition of nanoparticles to fluid changes the flow structure so that besides the increase in thermal conductivity, dispersion and fluctuation of nanoparticles, especially near the tube wall, lead to the increase in the energy exchange rates and augment the heat transfer rate between the fluid and the tube wall [22,23]. Moreover, the local Nu in fluid flow inside channel is related to the thickness of the thermal boundary layer, and a decrease in thermal boundary-layer thickness increases local Nu. One of the possible mechanisms responsible for the exhibition of the thermal boundary-layer thickness decrement by nanofluid is the migration of the nanoparticles due to shear action, Brownian motion, and the viscosity gradient in the cross section of the channel [47].

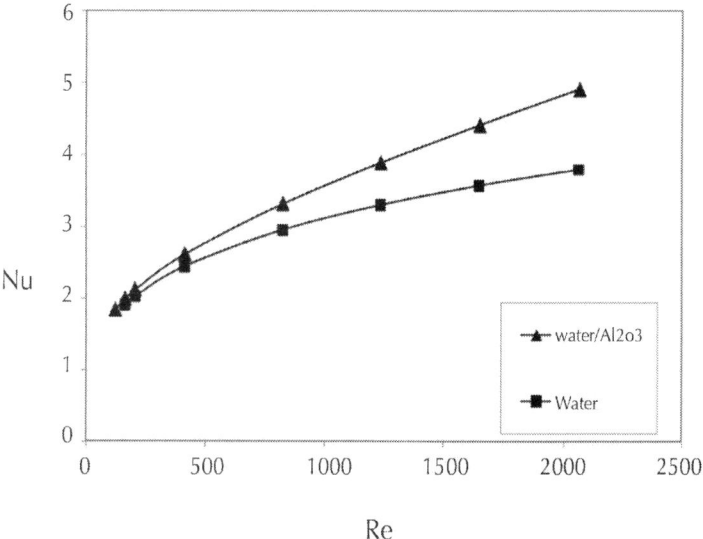

Figure 4: Comparison between nanofluid and pure fluid heat transfer.

Figure 5 shows the plots of the average Nusselt number versus *Re* at various concentrations of Al_2O_3 for 10-50 nm nanoparticles. This figure indicates that the average Nusselt number increases with the concentration of the nanoparticles, and better enhancement is seen at higher Reynolds numbers. For example, at d*p* = 10 and *Re* = 400, by increasing nanoparticle's concentration from 0.01 to 0.04, the average Nusselt number increases from 2.588 to 3.345, or at higher Reynolds number (*Re* = 2050), the Nusselt number changes from 4.89 to 6.02. The average Nusselt number at the same diameter increases according to Reynolds number. The results illustrate that by increasing Reynolds number from 500 to 2070 at d*p* = 30 nm, for 0.01 volume concentration of Al_2O_3/water nanofluid, the average Nusselt number increases from 2.57 to 4.53.

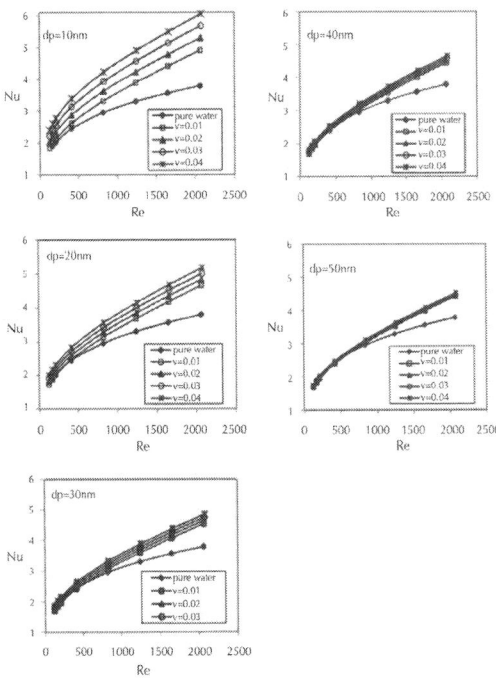

Figure 5: The influence of Al_2O_3 nanoparticles' volume concentration on the Nusselt number over a range of Reynolds numbers with diameter of nanofluids in the range of 10-50 nm.

During the nanofluid flow through the channel, migration of the nanoparticles and clustering due to non-uniform shear rate across the channel's cross section affect the heat transfer performance. Taking into account the increase in thermal conductivity of the nanofluid, other factors such as chaotic movement of nanoparticles, Brownian motion, and particles' migration must also be considered in the interpretation of heat transfer performance of nanofluids [35]. An increase in the volume fraction of the nanoparticles intensifies the interaction and collision of the nanoparticles. Also, diffusion and relative movement of these particles near the channel walls lead to the rapid heat transfer from the walls to the nanofluid. In other words, increasing the concentration of the nanoparticles intensifies the mechanisms responsible for the enhanced heat transfer. Moreover, at high flow rates, the dispersion effects and chaotic movement of the nanoparticles intensify the mixing fluctuations and change the temperature profile to a flatter profile similar to turbulent flow and cause an increase in the heat-transfer coefficient. At low flow rates, clustering and agglomeration of nanoparticles may exist in the nanofluid flow, and therefore, at a low Re, a lower heat transfer enhancement can be observed.

Figure 6 displays the effect of nanoparticle's diameter on the Nusselt number for Al_2O_3 nanofluids of constant volume concentrations. It can be seen that the average Nusselt number increases with the decreasing size of nanoparticles at the same concentration, particularly at high concentrations. For example, by increasing the size of the nanoparticles from 10 to 50 nm in 0.02 concentration atRe = 400, the average Nusselt numbers decrease from 2.845 to 2.376. Also, at Reynolds number 2050 in 0.02 volume concentration, increasing the nanoparticle's size from 10 to 50 nm leads to a decrease in Nu from 5.273 to 4.406. Similar kind of enhancement in Nusselt number with smaller particle size was observed from the experiments conducted by Zeinali Heris et al. [10].

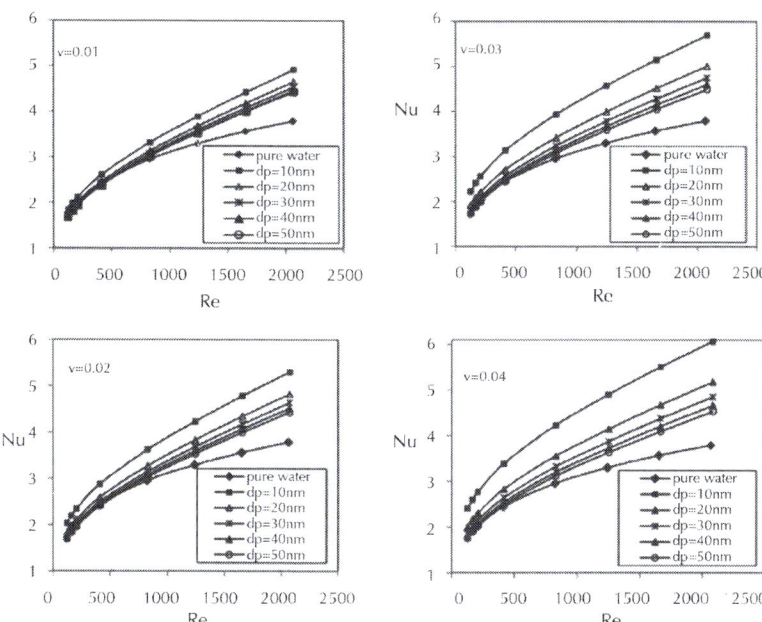

Figure 6: Effect of nanoparticle's diameter on the Nusselt number for. (a) 1% volume concentration, (b) 2% volume concentration, (c) 3% volume concentration, and (d) 4% volume concentration of Al_2O_3 nanofluids.

Since heat transfer between the nanoparticles and the fluid takes place at the particle-fluid interface, the ratio of the surface area of nanoparticles to their volume is the most important factor for heat transfer enhancement by nanofluids. On decreasing the size of nanoparticles, the ratio of surface area-to-volume of nanoparticles increases, allowing them to absorb and transfer heat more efficiently. Moreover, for the particles with very small diameter, the particles' distribution is fairly uniform; on the other hand, by increasing the nanoparticles' mean diameter, non-uniformity on the particles' distribution becomes more important, and the nanoparticles' concentration becomes higher at the vicinity of wall for which the viscous forces are important; this causes a decrement in the heat transfer enhancement while using the nanofluid with large particle's size.

For laminar flow, the heat transfer coefficient is mainly proportional to the fluid's thermal conductivity. Also, as already mentioned, the main effects of nanoparticles inside the fluid considering the Brownian motion and fluctuation are the change in the flow structure of fluid to semi-turbulence regime, and flattening of the transverse temperature gradient in the bulk of the fluid [9], and hence enhancing convective heat transfer of the nanofluid. However, in the turbulent regime, this mechanism is not dominant, and the thermal conductivity increment is the only factor for heat transfer enhancement in turbulent flow. For the turbulent flow, as the heat transfer coefficient depends to a smaller degree on the thermal conductivity of the fluid, the effect of thermal conductivity becomes less pronounced, since, in the presence of the nanoparticles, the aforementioned heat transfer enhancement may be decreased for turbulent flow.

A detailed review of the literature revealed that it might be beneficial to avail the advantage (i.e., lower pressure drop) of using triangular cross-sectional ducts in thermal engineering systems. However, heating exchange rates will decrease through such conduits. On the contrary, the results of this preliminary numerical study revealed that this could be compensated by using nanofluids in these systems, and so this will enhance the heat transfer rates. The use of nano-sized solid particles additives suspended into the base fluid (nanofluids) is a technique recommended for the enhancement of heat transfer in the triangular ducts. There are very few correlations available to exactly predict the heat transfer performance of nanofluids, as well as correlations which include the effects of solid particles' concentration, shape, size, dispersion, and nanoparticles' random movement are not suffice. Therefore, further research on convective heat transfer of nanofluids, and more theoretical and experimental research studies are needed to clearly understand and accurately predict their hydrodynamic and thermal characteristics especially in triangular ducts.

CONCLUSIONS

In this article, the laminar flow-forced convection of Al_2O_3/water nanofluid in a triangular duct is studied numerically. The results indicate that the addition of nanoparticles to base fluid, besides the thermal conductivity increment, affects the structure of the flow field and leads to heat transfer enhancement, because of the dispersion and random movement of nanoparticles inside the fluid. The results obtained by the numerical solutions show that decreasing the nanoparticle's size increases Nusselt number at a specific concentration, and increasing the nanoparticles' concentration increases Nusselt number at constant particle size. The results obtained in this preliminary study indicate that, in the case of using triangular cross-sectional ducts in thermal engineering systems, because of their low pressure drop, the decrement of heating exchange rate could be compensated by the use of nanofluids in these systems. Consequently, the flow of the nanofluids through triangular conduits has both the benefits of low pressure drop and high heat transfer rate.

AUTHORS' CONTRIBUTIONS

SZH planned the numerical investigation, took major part in the interpretation of results and participated in the manuscript preparation, SHN participated in the design of the study, ET drafted the manuscript, took part in the interpretation of results and participated in the manuscript preparation, and JS participated in the sequence alignment. All authors have read and approved the final manuscript.

ACKNOWLEDGEMENTS

The authors would like to thank the Nano Research Center of Iran for their financial support for this project.

REFERENCES

1.   Tauscher R, Mayinger F: Heat transfer enhancement in a plate heat exchanger with rib-roughened surfaces. Lehrstuhl afur Thermodynamik Technische Universitat Muchen, 85747 Garching, Germany; 1998.

2.   Sahin AZ: Irreversibility's in various duct geometries with constant wall heat flux and laminar flow. Energy 1998, 23(6):465-473.

3.   Kakac S, Bergles AE, Mayinger F: Heat Exchangers. Thermal-Hydraulic Fundamentals and Design New York: McGraw-Hill; 1981.

4.   Namburu PK, Das DK, Tanguturi KM, Vajjha RS: Numerical study of turbulent flow and heat transfer characteristics of nanofluids considering variable properties. Int J Thermal Sci 2009, 48:290-302.

5.   Choi SUS: Enhancing thermal conductivity of fluid with nanoparticles. In Developments and Application of Non-Newtonian Flows. Volume 66. Edited by: Siginer DA Wang HP. New York: ASME; 1995:99-105.

6.   Lee S, Choi SUS, Li S, Eastman JA: Measuring thermal conductivity of fluids containing oxide nanoparticles. J Heat Transf 1999, 121:280-289.

7.   Pak BC, Cho YI: Hydrodynamic and heat transfer study of dispersed fluids with submicron metallic oxide particles. Exp Heat Transf 1999, 11:151-170.

8.   Lee S, Choi SUS: Application of metallic nanoparticle suspensions in advanced cooling systems. Proceeding of International Mechanical Engineering Congress and Exposition Atlanta, USA; 1996.

9.   Xuan Y, Li Q: Investigation on convective heat transfer and flow features of nanofluids. J Heat Transf 2003, 125:151-155.

10. Zeinali Heris S, Etemad SGh, Nasr Esfahany M: Experimental investigation of oxide nanofluids laminar flow convective heat transfer. Int Commun Heat Mass Transf 2006, 33:529-533.

11. Zeinali Heris S, Nasr Esfahany M, Etemad SGh: Investigation of CuO/water nanofluid laminar convective heat transfer through a circular tube. J Enhanc Heat Transf 2006, 13(4):1-11.

12. Zeinali Heris S, Nasr Esfahany M, Etemad SGh: Experimental investigation of convective heat transfer of Al2O3/water nanofluid in circular tube. Int J Heat Fluid Flow 2007, 28:203-210.

13. Maiga SB, Nguyen CT, Galanis N, Roy G, Mare T, Coqueux M: Heat transfer enhancement in turbulent tube flow using Al2O3 nanoparticle suspension. Int J Numer Methods Heat Fluid Flow 2006, 16:275-292.

14. Akbari M, Behzadmehr A: Developing mixed convection of a nanofluid in a horizontal tube with uniform heat flux. Int J Numer Methods Heat Fluid Flow 2007, 17:566-586.

15. Das MK, Ohal PS: Natural convection heat transfer augmentation in a partially heated and partially cooled square cavity utilizing nanofluids. Int J Numer Methods Heat Fluid Flow 2009, 19:411-431.

16. Ben Mansour R, Galanis N, Nguyen CT: Developing laminar mixed convection of nanofluids in an inclined tube with uniform wall heat flux. Int J Numer Methods Heat Fluid Flow 2009, 19:146-164.

17. Zeinali Heris S, Etemad SGh, Nasr Esfahany M: Convective heat transfer of a Cu/water nanofluid flowing through a circular tube. Exp Heat Transf 2009, 22:217-227.

18. Das SK, Putra K, Roetzel W: Pool boiling characteristics of nano-fluids. Int J Heat Mass Transf 2003, 46(5):851-862.

19. Soltani S, Etemad SGh, Thibault J: Pool boiling heat transfer performance of Newtonian nanofluids. Heat Mass Transf 2009, 45(12):1555-1560.

20. Vassallo P, Kumar R, Amico SD: Pool boiling heat transfer experiments in silica-water nanofluids. Int J Heat Mass Transf 2004, 47:407-411.

21. Bang IC, Chang SH: Boiling heat transfer performance and phenomena of Al2O3/water nanofluids from a plain surface in a pool. Int J Heat Mass Transf 2005, 48:2407-2419.

22. Noie SH, Zeinali Heris S, Kahani M, Nowee SM: Heat transfer enhancement using Al2O3/water nanofluid in a two-phase closed thermosyphon. Int J Heat Fluid Flow 2009, 30:700-705.

23. Hamed Mosavian MT, Zeinali Heris S, Etemad SGh, Nasr Esfahany M: Heat transfer enhancement by application of nano-powder. J Nanoparticle Res 2010, 12:2611-2619.

24. Bergles AE: Heat transfer enhancement-the encouragement and accommodation of high heat fluxes. J Heat Transf 1997, 119:8-19.

25. Kays WM, London AL: Compact Heat Exchangers New York: McGraw-Hill; 1984.

26. Eckert ERG, Irvine TF, Yen JT: Laminar heat transfer in wedge-shaped passage. Trans ASME 1958, 80:1433-1438.

27. Sparrow EM: Laminar flow in isosceles triangular ducts. AICHE J 1962, 5:599-604.

28. Sparrow EM, Haji-sheikh A: Laminar heat transfer and pressure drop in isosceles triangular, right triangular and circular sector ducts. ASME J Heat Transf 1964, 87:426-427.

29. Schmidt FW, Newell ME: Heat transfer in fully developed flow through rectangular and isosceles triangular ducts. Int J Heat Mass Transf 1967, 10:1121-1123.

30. Shah RK: Laminar flow friction & forced convection heat transfer in ducts of arbitrary geometry. Int J Heat Mass Transf 1975, 18:849-862.

31. Shah RK, London AL: Laminar Flow Forced Convection in Ducts New York: Academic Press Inc; 1978.

32. Zhang LZ: Laminar flow and heat transfer in plate-fin triangular ducts in thermally developing entry region. Int J Heat Mass Transf 2007, 50:1637-1640.

33. Gupta RV, Geyer PE, Fletcher DF, Haynes BS: Thermohydraulic performance of a periodic trapezoidal channel with a triangular cross-section. Int J Heat Mass Transf 2008, 51:2925-2929.

34. Kuznetsov AV, Cheng L, Xiong M: Effects of thermal dispersion and turbulence in forced convection in a composite parallel-plate channel: investigation of constant wall heat flux and constant wall temperature cases. Numer Heat Transf A 2002, 42:365-383.

35. Zeinali Heris S, Nasr Esfahany M, Etemad SGh: Numerical investigation of nanofluid laminar convective heat transfer through circular tube. Numer Heat Transf A 2007, 52:1043-1058.

36. Xuan Y, Rotzel W: Conception for heat transfer correlation of nanofluid. Int J Heat Mass Transf 2000, 43:3701-3708.

37. Drew DA, Passman SL: Theory of Multi Component Fluids Berlin: Springer; 1999.

38. Kaviany M: Principles of Heat Transfer in Porous Media Berlin: Springer; 1995.

39. Taylor GI: Dispersion of soluble matter in solvent flowing through a tube. Proc R Soc Lond 1954, A21:186.

40. Khaled A-RA, Vafai K: Heat transfer enhancement through control of thermal dispersion effects. Int J Heat Mass Transf 2005, 48:2172-2185.

41. Akbarinia A, Behzadmehr A: Numerical study of laminar mixed convection of a nanofluid in horizontal curved tubes. Appl Therm Eng 2007, 27:1327-1337.

42. Eastman JA, Choi SUS, Li S, Yu W, and Thomson LJ: Anomalously increased effective thermal conductivities of ethylene glycol-based nanofluids containing copper nanoparticles. Appl Phys Lett 2001, 78:718-720.

43. Wang XQ, Mujumdar AS: A Review on nanofluids-part II: experiments and applications. Brazilian J Chem Eng 2008, 25(4):631-648.

44. Yu W, Choi SUS: The role of interfacial layers in the enhanced thermal conductivity of nanofluids: a renovated Maxwell model. J Nanoparticle Res 2003, 5:167-171.

45. Jang SP, Choi SUS: Role of Brownian motion in the enhanced thermal conductivity of nanofluids. Appl Phys Lett 2004, 84:4316-4318.

46. Naris S, Valougeorgis D: Rarefied gas flow in a triangular duct based on a boundary fitted lattice. Eur J Mech B 2008, 27:810-822.

47. Wen D, Ding Y: Experimental investigation into convective heat transfer of nanofluid at the entrance rejoin under laminar flow conditions. Int J Heat Transf 2004, 47:5181-5188

Index